RECENT ADVANCES
AND ISSUES IN
Molecular
Nanotechnology

Frontiers of Science Series

RECENT ADVANCES AND ISSUES IN Molecular Nanotechnology

By David E. Newton

An Oryx Book

GREENWOOD PRESS
Westport, Connecticut • London

Library of Congress Cataloging-in-Publication Data

Newton, David E.
    Recent advances and issues in molecular nanotechnology/David E. Newton.
       p. cm.—(Frontiers of science series)
    Includes bibliographical references and index.
    ISBN 1-57356-307-2 (alk. paper)
    1. Nanotechnology.   I. Title.   II. Series.

T174.7.N49 2002
620'.5—dc21   2002022444

British Library Cataloguing in Publication Data is available.

Library of Congress Catalog Card Number: 2002022444
ISBN: 1-57356-307-2

First published in 2002

Greenwood Press, 88 Post Road West, Westport, CT 06881
An imprint of Greenwood Publishing Group, Inc.
www.greenwood.com

Printed in the United States of America

The paper used in this book complies with the
Permanent Paper Standard issued by the National
Information Standards Organization (Z39.48-1984).

10 9 8 7 6 5 4 3 2 1

To Dan Trevalle
who has made so much possible for us,
and with such charm, grace, and efficiency

Contents

Preface

The term *molecular nanotechnology* is probably unfamiliar to most people in the world today. Yet the form of research designated by this phrase could have enormous potential in reshaping human civilization. Some scientists believe that developments in the field could lead to a vastly increased human life span, computing devices of unparalleled speed and efficiency, and materials with properties not found in any natural or synthetic product yet known. Such developments could also lead to social, political, and economic challenges the scope of which the human race has never seen.

Molecular nanotechnology refers to the manipulation of individual atoms and molecules so as to produce materials with precise, predictable properties. The possibility of such research was first suggested more than 50 years ago by Nobel Laureate Richard Feynman. At the time of Feynman's speech on the topic, no scientist had the slightest inkling as to how his ideas could be realized. Many scientists today are hesitant as to how far such research can develop in the production of usable products for human civilization. The idea of building designed matter an atom or a molecule at a time still strikes some scientists and nonscientists alike as science fiction, a fantasy that can only distract researchers from more realistic and more productive forms of work.

The manipulation of atoms and molecules is no longer a fantastic dream, however. Researchers throughout the world are carrying out such projects every day. As a result of that work, scientists are beginning to have a glimpse of a coming revolution in their field, a revolution in which the most unlikely devices and processes seem to have some promise of realization. As to whether molecular nanotechnology will ever reach the point where it realizes the greatest hopes of its most optimistic proponents is still uncertain. But the road that leads in that direction is an exciting one that will surely produce products that will change the everyday lives of people throughout the world.

Introduction

One of the most important directions of scientific research in the last decade has been miniaturization. Scientists are looking for new ways to work with matter at smaller and smaller dimensions. At one time, an ordinary optical microscope was sufficient for the most detailed study that any scientist might undertake. Hardly anyone was interested in examining objects with dimensions of less than a few micrometers (millionths of a meter). Gradually, however, it has become necessary to explore regions even smaller than the microscopic, regions where the dimensions of objects are measured in a few nanometers (billionths of a meter).

Perhaps the most important force in the drive for miniaturization has been the demands of computer technology. Progress in the computer industry means building faster, more efficient machines, which, in turn, require the manufacture of smaller and smaller units. At one time, the manufacture of a circuit board could be carried out with delicate, hand-held tools. The demand for miniaturization, however, has led to the development of far more sophisticated technology, such as the laser beams and high-energy electron beams now used for the etching of silicon wafers employed in the manufacture of circuit boards.

For some time, computer scientists have realized that even the best technology available today would be useful for only a few years more. To make even smaller computer components, an entirely new technology would have to be developed. The manipulation of individual atoms and molecules—an approach that we call *molecular nanotechnology* here—holds the potential to become that new technology.

This book provides basic information on this revolutionary new form of technology. It makes little or no attempt to review the vast amount of research and development still being done with more conventional forms of technology designed for use at the nanoscale level. That research is important and interesting and worthy of the reader's attention. We focus, however, on the newer technologies based on the manipulation of individual atoms and molecules.

Chapters 1 and 2 discuss the nature of nanotechnology in general and explains the differences between "top-down" and "bottom-up" approaches. Special attention is paid to the ideas of Nobel Laureate Richard Feynman and K. Eric Drexler, one of the earliest proponents of "bottom-up" nanotechnology. Drexler's ideas continue to be the source of great controversy within the scientific community; some see him as a prophet of a new age in science, whereas others regard him as an incompetent dreamer. Whatever the fate of his ideas, Drexler has fashioned a new paradigm for the way in which some forms of scientific research might be conducted. His work provides a general structure within which to view much of the most exciting research now going on.

Chapter 3 focuses on this research. It outlines some of the most interesting discoveries in the field of protein and nucleic acid chemistry, carbon nanotubes, molecular electronics, and nanoscale devices and processes.

Chapter 4 is devoted to brief biographical sketches of some important individuals in the field. Chapter 5 provides a chronology of important events. Chapter 6 includes some important documents relating to molecular nanotechnology, ranging from early papers by Feynman and Drexler to recent governmental initiatives on nanotechnology.

Chapters 7 through 10 provide information, data, and listings of associations, organizations, corporations, and academic institutions involved in research related to molecular nanotechnology. Chapters 11 and 12 provide print and nonprint resources, respectively, in the area of molecular nanotechnology. Finally, Chapter 13 is a glossary of commonly used terms in the area of molecular nanotechnology.

Chapter One

Molecular Nanotechnology Today

A revolution is under way in the world today of which relatively few people are aware. That revolution involves the field known as *nanotechnology*. Many scientists believe that developments resulting from research in nanotechnology will entirely reshape the nature of human civilization, with positive and negative impacts. For example, advances in nanotechnology may eliminate disease entirely and extend the normal human life span to well over 100 years. That research may, however, also result in new products that could bring about the end of human life on Earth.

Nanosizes

The term *nanotechnology* comes from the Greek prefix -*nano*, meaning "dwarf." In the metric system, the prefix -nano means "one billionth." A nanometer (abbreviation, *nm*) is one billionth (10^{-9}) of a meter, and a nanosecond (*ns*) is one billionth of a second.

To appreciate the size of a nanometer, consider that a human hair is about 10,000 nm in diameter. The nanometer-scale world is that of atoms and molecules. For example, the smallest of all atoms, the hydrogen atom, has a diameter of 0.078 nm. Most molecules are larger than an individual atom. Protein and nucleic acid molecules are typically composed of many thousands of atoms with dimensions of tens of nanometers or more.

Nanotechnology and Microtechnology

Over the past two decades, the number of references to nanotechnology in the scientific literature has increased significantly. The term is often used ambiguously, however, having somewhat different meanings depending on the context in which it appears.

In many cases, the term is used simply to describe objects and events that occur within "very small" dimensions. Those dimensions may be at the nanometer level, or they may be at a larger dimension, such as that of a few microns. A *micron* (abbreviation, μm) is a micrometer, or one millionth (10^{-6}) of a meter. Inarguably, a micron is a very small dimension. The width of a human hair is about 10 μm. Still, a micron is 1,000 times larger than a nanometer. The difference between a nanometer and a micron is similar to the difference between a ladybug and an elephant.

Some important research now being conducted could be labeled as *microtechnology* because it involves objects and events with dimensions in the micron range. For example, it is now possible to use laser beams to etch lines only a few nanometers wide on the surface of a material. Also, "ion guns" can be used to accelerate charged particles toward the surface of a material, where they are deposited and used to construct various shapes and structures with micron or submicron dimensions.

Authors sometimes refer to such research as "nanotechnology" because the dimensions involved seem truly "very small." In any given case, however, the dimensions involved may range from a few nanometers to a few microns. The distinction as to what constitutes "microtechnology" and what constitutes "nanotechnology" may sometimes become blurred. This distinction is far more than a semantic difference of opinion, however. As used by many people today, the term *nanotechnology* refers not simply to an order of magnitude, but to an entirely new way of working with materials.

For all of human history, new products have been made by starting out with a hunk of metal, stone, wood, or other material and hacking away at it to get just the right size and shape. At the earliest stages, humans chipped away on bone, wood, and other materials with hard rock to form arrowheads and spearheads. Today, complex machinery is used to bend, twist, cut, meld, and otherwise shape aluminum, steel, and other materials into automobile bodies, skyscraper skeletons, refrigerators, and other objects. These techniques are sometimes referred to as "saw and polish" or "shake and bake" approaches to the manipulation of matter.

As different as primitive and modern technologies may seem, they both depend on a "top-down" approach to the manufacture of objects. One begins with a mass of trillions and trillions of atoms and molecules and with brute force arranges those atoms and molecules into some desirable form. Without exception, this approach produces imperfect products, with cracks, holes, and other defects that eventually result in the failure of the product.

One form of nanotechnology operates on a totally different philosophy, one that could be called the "bottom-up" approach. In this approach, atoms and molecules are assembled one at a time in some predetermined and desired manner. Every particle goes into exactly the right position such that the final product will be an automobile frame, an I-beam, or a toaster with a completely perfect molecular composition. The product theoretically will have no flaws and will last many times longer than any product currently available.

Proponents of this approach to the manipulation of matter have suggested a solution for the semantic problem of the term *nanotechnology*. They have suggested that this research be called *molecular nanotechnology*, leaving free the more general term of *nanotechnology* for studies at the "very small" range. Many other terms have been suggested for the "bottom-up" approach, including *molecular manufacturing*, *molecular fabrication*, *mechanosynthesis*, and *chemosynthesis*. The term *mesotechnology* ("middle technology") has also been proposed to describe research at dimensions of less than the micron level (microtechnology) and greater than the atomic or nanometer level (molecular nanotechnology).

As just described, molecular nanotechnology may sound a bit like science fiction. It requires that humans manipulate with perfection the tiniest particles of which matter is composed, particles that seem to have no definite shape and that may follow the rules of quantum mechanics rather than those of everyday Newtonian physics. It is hardly surprising that many critics have rejected the notion of molecular nanotechnology as a viable field of scientific research. What may be surprising is that the theoretical and experimental basis for molecular nanotechnology is now established so firmly that academic, governmental, and industrial laboratories all over the world are now actively engaged in research in molecular nanotechnology. Some reputable organizations and individuals now seem to be suggesting that molecular nanotechnology not only may be possible but perhaps inevitable. It is no longer unusual or surprising to find optimistic statements such as the following in the scientific and popular literature: "It no longer seems a question of whether nanotechnology will become a reality. The big questions are how important and transformative nanotechnology will become, will it become affordable, who will be the leaders, and how can it be used to make the world a better place?" ("Nanotechnology: Shaping the World Atom by Atom," 8).

Origins of Molecular Nanotechnology

The birth of molecular nanotechnology is usually traced to a speech made by Nobel Laureate Richard Feynman at a meeting of the American Physical Society in December 1959. Feynman's speech was entitled "There's Plenty of Room at the Bottom." In his remarks, Feynman outlined the continuing challenge facing researchers who are designing and manufacturing smaller and smaller objects.

Feynman first suggested some fairly traditional methods for continually reducing the dimensions at which research is carried out. He described machines that were programmed to make exact copies of themselves, but at one half their original dimensions. He then went on to propose an entirely new approach to research at the smallest possible dimensions. Why not, he said, find ways to move individual atoms and molecules about, arranging them in exactly the shapes needed for any given object? He told his listeners that "The principles of physics, as far as I can see, do not speak against the possibility of maneuvering things atom by atom. . . . Put the atoms down where the chemists says, and so you make the substance" (Feynman, 1960, 24).

At the conclusion of his speech, Feynman proposed two challenges to his audience. He offered two cash prizes in the amount of $1,000 each for the first person to (1) build an operating electric motor with cubic dimensions of no more than 1/64 inch on a side and (2) find a way to reproduce all of the printed information on a page by 1/25,000 so that it could be read with an electron microscope. The first challenge was answered (by "top-down" technology) within a year of Feynman's speech; the second was not answered until 26 years later.

Feynman's speech had remarkably little impact on his listeners or on the physics community as a whole. At the time, most chemists and physicists thought of atoms and molecules as blurry little particles without exact dimensions and, perhaps, without much physical reality. They were usually described in the form of mathematical equations that provided little guidance as to how they could be "picked up and moved around." In addition, most chemists and physicists thought that certain physical laws, such as the uncertainty principle and limitations imposed by thermal motion, would make it impossible to carry out the kind of research Feynman was suggesting.

Adding to his listeners' uncertainties was Feynman's own reputation as someone with an active, off-beat sense of humor. (One of Feynman's most popular works was a book entitled *Surely You're Joking, Dr. Feynman.*) Was Feynman serious in his proposals, his colleagues wondered, or was his speech just one grand practical joke? Feynman tried to make clear how serious he was about his proposals. "I am not inventing anti-gravity, which is possible someday only if the laws are not what we think," he said. "I am telling you what could be done if the laws are what we think; we are not doing it simply because we haven't yet gotten around to it" (Feynman, 1960, 24).

Feynman's proposals languished for more than two decades, the victim of massive disinterest on the part of the scientific community. Finally, in the early 1980s, the idea of atomic and molecular manipulation and construction was reinvented, not by a Nobel Laureate, but by an undergraduate at the Massachusetts Institute of Technology (MIT).

K. Eric Drexler and Molecular Nanotechnology

It is rare in the history of science that one can point to a single individual as the "creator" or "inventor" of a new field of science. Molecular nanotechnology is one of the few instances in which that can be said. The individual responsible for the creation of molecular technology is K. Eric Drexler.

Drexler entered MIT as a freshman in fall 1973. At the time, he was interested primarily in space exploration and already was engaged in the design of components to be used for human space colonies. During his first year at MIT, he met with Gerard K. O'Neil, then at Princeton University and one of the leading proponents of space colonization by humans. Drexler and O'Neil found that they had a great deal in common and forged a friendship that was to last many years.

Before long, Drexler also became interested in another totally different field of research: genetic engineering. In the early 1970s, biochemists were just beginning to discover methods for manipulating molecules of deoxyribonucleic acid (DNA), the molecular "master plan" that controls the activity of all living cells. Drexler read about this research and decided that the general principles of genetic engineering might also be applicable to other kinds of molecules. He began to consider the possibility of putting together any structure at all, one atom or one molecule at a time, "from the bottom up." That is, he reinvented the concept of molecular nanotechnology originally proposed by Richard Feynman two decades earlier.

Drexler had never heard about Feynman's speech. It was not until his then-girlfriend and wife-to-be Chris Peterson brought him an article in the November 1979 issue of *Physics Today* that mentioned Feynman's talk that he learned of the Nobel Laureate's ideas. Peterson was also responsible for bringing to Drexler's attention a second important article only a few months later. In that article, "Protein Macromolecules Interface to Microcircuitry" (*Semiconductor International*, May 1980, 10), engineer James McAlear pointed out the possibility of using "genetic engineering to produce protein molecules for the specific application to microdevice fabrication" (as quoted in Regis, 1995, 77). But that was just the idea that Drexler had had at least three years earlier.

Drexler realized that it was time for him to put his ideas into a formal paper. That paper, "Molecular Engineering: An Approach to the Development of General Capabilities for Molecular Manipulation," was published in the *Proceedings of the National Academy of Sciences* in 1981. The paper outlined most of the fundamental concepts on which molecular nanotechnology is based. Drexler later wrote a more extended popular version of the paper, *Engines of Creation: The Coming Era of Nanotechnology*, a book published in 1986.

By the time *Engines of Creation* had appeared, Drexler's life had changed considerably. He had received his bachelor's and master's degrees from MIT in 1977 and 1979, moved to California in 1985, helped to found the Foresight Institute for the study of molecular nanotechnology, and was preparing to teach the first course in the subject of molecular nanotechnology at Stanford University.

General Principles of Drexlerian Molecular Nanotechnology

The fundamental concepts on which Drexler's vision of molecular nanotechnology is based are relatively simple. They posit the need for three basic structures: (1) assemblers and disassemblers, (2) replicators, and (3) nanocomputers.

Assemblers and Disassemblers

An assembler is a device for carrying out some mechanical action, such as picking up an atom and moving it to another position. Disassemblers are basically the same as assemblers except that their job is to take things apart rather than put them together.

Assemblers can assume many different shapes, depending on the function they are expected to perform. Most engineers would recognize the various types and components of assemblers from their macroscopic counterparts. Some examples of assemblers and the parts of which they are made are pumps, bearings, cables, struts and beams, fasteners, drive shafts, gears, clamps, conveyor belts, motors, and containers. All of these devices are used in the everyday world to make, hold, and maneuver objects and materials. The only difference in such devices designed for the nanolevel is size: They consist of a few thousand or a few million atoms or molecules rather than the trillions and trillions of particles contained in everyday objects of the same kind.

An example of a planetary gear for use in molecular nanotechnology is shown in Figure 1.1. The device would perform the same functions on the nanolevel that familiar planetary gears perform on the macroscopic level. The only difference is that the planetary gear shown here contains only 3,557 atoms.

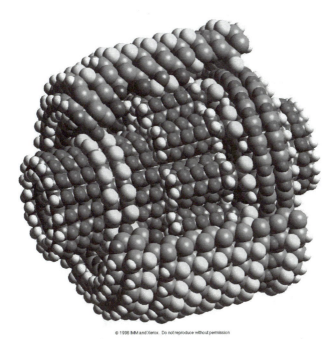

Figure 1.1. Computer model of a planetary gear nanodevice.
© *Institute for Molecular Manufacturing (IMM)*
<*www.imm.org*>. *All rights reserved.*

The availability of a wide variety of assemblers would make possible the construction of any material by moving individual atoms and molecules according to any predetermined plan. Objects constructed by this method would have perfect structures in the sense that every atom or molecule is in exactly the correct place. No flaws would exist, and the object would be much stronger and longer lasting than objects built by traditional "top-down" methods.

Replicators and Nanocomputers

Replicators are specialized assemblers programmed to make exact copies of themselves. Combinations of assemblers will be available that will allow the construction of any object or material that can be imagined. One possibility is to program assemblers to make other objects exactly like themselves. Assemblers with this capability are called *replicators*.

The idea of self-replicating devices at the molecular level is hardly a new one. DNA, ribonucleic acid (RNA), and protein molecules work together in living cells to accomplish the same objective. The construction of self-replicating machines had been the subject of serious study among scientists for many decades. The only new factor in such studies is the reduction of the scale at which such self-replication takes place to the nanometer level.

Nanocomputers are devices that carry the instructions for the operation of assemblers, disassemblers, and replicators. As with these three types of machines, nanocomputers will be constructed and will operate on the nanometer scale. At this time, it is not clear whether nanocomputers will operate on electronic or mechanical principles. Electronic computers would operate faster than mechanical computers. It is not clear, however, whether the technology used to make electronic computers can be scaled down to the nanometer level without introducing new problems. The phenomenon known as *tunneling*, for example, may introduce significant new problems for the use of electrons in carrying messages in a computer.

The alternative to electronic computers is mechanical computers based on the broad general design first suggested in the mid-1800s by Charles Babbage. Mechanical computers would operate more slowly than electronic computers, but they have the advantage of being much smaller than the smallest electronic computer that can now be imagined. As a result, messages in a mechanical computer will travel more slowly than those in an electronic computer, but they will travel over much smaller distances.

Responses to Drexler's Ideas

At first, Drexler's *Engines of Creation* drew slightly more response than had Richard Feynman's 1959 speech before the American Physical Society. It was reviewed favorably in *The New York Times* and *Technology Review*, a magazine for general readers published by MIT. The book was largely ignored by other specialized and general journals in science.

By 1990, however, some scientists and science journalists were beginning to look more carefully at Drexler's ideas for nanoscale machines. In many cases, they raised serious doubts as to the feasability of those ideas. In 1990, journalist Simon Garfinkel wrote in the Summer 1990 issue of *Whole Earth Review* about what he called the "cult of nanotechnology." Molecular nanotechnology, Garfinkel said, envisions "working with atoms the same way a model-maker might work with wooden sticks and styrofoam balls." But the problem, Garfinkel explained, is that "atoms don't work that way" (Garfinkel, 1990, 105).

To support his view, Garfinkel cited scientists he had interviewed for the article, including Robert J. Silby and Rick L. Danheiser, both professors of chemistry at MIT, and James S. Nowick, a graduate student in organic chemistry at MIT. The problem with Drexler's ideas, Silby pointed out, is that they consider molecular systems as mechanical systems, and the two are not comparable. "Molecules are not rigid," Silby said, "they vibrate, they have bending motions" (Garfinkel, 1990, 105). Nowick's main criticism was that Drexler was "coming forth as a visionary without actually doing anything" (Garfinkel, 1990, 107). Garfinkel's own conclusion was that some form of nanotechnology would develop in the future, but that it would be "a far cry from self-reproducing, self-repairing Nanomachines driven by tiny mechanical computers" (Garfinkel, 1990, 113). The *Whole Earth Review* article is an especially interesting look at early comments about molecular nanotechnology because it also provides Drexler's response to Garfinkel's article, then a counterresponse by Garfinkel and a final response by Drexler.

Drexler has had more than his share of critics over the succeeding years. The alternative paper *SF Weekly* published an extensive article in late 1999 about his ideas and the views of Bay Area scientists about the possibilities of molecular nanotechnology. The author, Peter Byrne, barely seemed able to control his mirth about the absurdity of Drexler's ideas. "The extraordinarily unlikely nanotech products envisioned by exaggeration-prone media outlets," he wrote, "range from molecular sensors in flimsy underwear that tell smart washing machines what water temperature they should use to artificial red blood cells to evil swarms of planet-devouring molecules" (Byrne, 1999, 16–17). He depicted Drexler's ideas as "long on imagination and short on facts" and "nano-pie-in-the-sky" (Byrne, 1999, 17). An interesting feature of the article is that some of the Bay Area researchers (including Hongjie Dai; see Chapter 3) have produced nanodevices of the type Drexler originally described in their own work.

The article questioning Drexler's ideas that has drawn the most attention is probably one that appeared in the popular magazine *Scientific American* in 1996. Author Gary Stix described Drexler as "the 40-year-old guru of the nanoists, [who] speaks with an exaggerated professional tone that is faintly reminiscent of the pedantic 1960s cartoon character Mr. Peabody." Stix characterized Drexler's ideas as "fanciful scenarios [that have] nonetheless come to represent nanotechnology for many aesthetes of science and technology." Like Byrne, Stix attributed the popularity of Drexler's ideas about molecular nanotechnology to a "public image . . . shaped by futurists, journalists and science-fiction scribes [which] contrasts with the reality of the often plodding and erratic path that investigators follow in the trenches of day-to-day laboratory research and experimentation" (Stix, 1996, 94).

Proponents of molecular nanotechnology reacted vigorously to the Stix article. Carl Feynman, son of Richard Feynman, wrote *Scientific American* to object to Stix's misuse of his father's name and the misrepresentation of his opinion about nanotechnology. Ralph Merkle, then at Xerox PARC, wrote an extended rebuttal to the article, pointing out that the only authority cited was Dr. David Jones, columnist for the prestigious journal *Nature*. A year previously, Jones had reviewed Ed Regis' book *Nano*, calling nanotechnology a field that "need not be taken seriously." He said that it is "just another exhibit in the freak show that is the boundless-optimism school of technical forecasting" (Jones, 1995, 837).

The exchange of viewpoints between *Scientific American* and members of the molecular nanotechnology community continued for nearly a year. A full report of that exchange, along with relevant documents from both sides, is available on the Foresight Institute's website at http://www.foresight.org/SciAmDebate/SciAmOverview.html. In the end, molecular nanotechnologists seemed to have been satisfied that the magazine had finally taken a more positive view toward their subject. The website report concludes with the Institute's comment: "We salute *Scientific American* for recognizing the growing importance of nanotechnology" ("Foresight Debate," 1997, 3).

Foresight's optimistic assessment was somewhat premature. In September 2001, *Scientific American* published a special issue on nanotechnology that included remarks that further questioned the feasibility of Drexlerian nanodevices. In his article, "The Once and Future Nanomachine," George Whitesides wrote that "The dream of the [Drexlerian] assembler holds seductive charm in that it appears to circumvent these myriad difficulties [of assembling on a molecular level]. This charm is illusory; it is more appealing as metaphor than as reality, and less the solution of a problem than the hope for a miracle."

In the same issue, Nobelist Richard Smalley also tried to dash cold water on the dreams of Drexlerian nanodevices. He wrote, "Self-replicating, mechanical nanorobots are simply not possible in our world." Smalley has also reiterated this view in other forums. For example, in March 2001, National Science Foundation publication *Societal Implications of Nanoscience and Nanotechnology*, he wrote:

> For fundamental reasons I am convinced that these nanobots are an impossible, childish fantasy. . . . We should not let this fuzzy-minded nightmare dream scare us away from nanotechnology. Nanobots are not real.

As in their previous debate with the magazine, Drexler's supporters responded with a vigorous campaign of letter- and article-writing aimed at rebutting the negative views of Whitesides and Smalley. Their strongest argument can be found on the Institute for Molecular Manufacturing's website, "A Debate about Assemblers," at http://www.imm.org/SciAmDebate2/.

The important point about this debate is that there remains a massive gulf between those who see something approaching an unbounded future for molecular nanotechnology, akin to the vision that Drexler produced two decades ago, and those who believe that fundamental physical principles place an irrevocable limit as to the accomplishments that nanotechnologists can expect to achieve. In a truly classical sense, "only time will tell" which of these positions is correct.

The objections raised by many prominent scientists to Drexler's ideas are hardly surprising. These ideas were and are profoundly different from the principles that drive nearly all scientific research today. In 1970, philosopher of science Thomas Kuhn wrote a historic book, *The Structure of Scientific Revolutions*, in which he analyzed the mindset of scientists faced with revolutionary new ideas in their fields. He argued that, in such cases, scientists tend to reject the new ideas out of hand, often without giving them a legitimate hearing. He suggested that new ideas in science are not gradually accepted by scientific communities. Instead, they are embraced only when an older generation of researchers retires or dies and a new generation, with a greater openness for change, becomes dominant in the field.

From Kuhn's point of view, opposition to molecular technology is not only understandable, but also entirely to be accepted. It is conceivable that many researchers who are currently critical of Drexler's ideas may be won over eventually by his arguments. It is also conceivable, however, that molecular nanotechnology will not become a widely accepted field of science until younger researchers committed to the field appear and become active in molecular nanotechnology research.

Drexler has long been aware of the seeming "science fiction" cast to his ideas about molecular nanotechnology and the resulting opposition those ideas were likely to encounter. In *Engines of Creation*, he analyzed this problem from a Kuhnian point of view. Instead of adopting Kuhn's analysis, however, he turned to a theory developed by zoologist Richard Dawkins in his book *The Selfish Gene* (New York: Oxford University Press, 1976). In that book, Dawkins argued that certain mental patterns (such as ideas) exist within a human population and are subjected to evolutionary tests (such as "survival of the fittest") at the intellectual level in much the way that genes are tested at the biological level. Dawkins called those mental patterns *memes*.

Notions as to what can or cannot be done in scientific research might be classified as memes. The meme of "top-down" technology has dominated human thought for millennia. Drexler's suggestions for a "bottom-up" technology have created a new meme that may have to compete with the older "top-down" meme. An intellectual battle between these two memes may be inevitable, although, as with most evolutionary competitions, one can hardly predict the outcome of this battle.

Uncertainty Principle

There are some legitimate and logical questions about molecular nanotechnology that would occur to anyone with a limited knowledge of chemistry and physics. Two of the most obvious problems involve the uncertainty principle and kinetic motion of particles.

The uncertainty principle was first enunciated by the German physicist Werner Heisenberg in 1927. According to this principle, it is impossible to know precisely the position and velocity of any particle simultaneously at any one moment in time. The product of the uncertainties for these two measurements, according to Heisenberg, is greater than or equal to a constant known as *Planck's constant*. Its value is 6.63×10^{-34} Joule seconds.

Simple calculations show that the uncertainty principle has no practical application to objects at the macroscopic or microscopic scale. It does have to be taken into consideration, however, when dealing with the smallest of particles, such as the electron. What the uncertainty principle says about electrons is that one can never be entirely sure about the location or movement of individual electrons at any given moment. It is for this reason that modern models of atoms always depict their electronic structures as "probability clouds," regions in which an electron is more or less likely to be found.

At first glance, it seems that the uncertainty principle might pose serious problems for molecular nanotechnology. Some researchers have expressed the view that it might be impossible to locate atoms precisely enough to pick them up and move them around, as would be necessary in building structure from the bottom up.

Such fears are probably unjustified. Although the uncertainty principle does restrict our knowledge about electrons, it has much less relevance to atoms and molecules as whole units. In such cases, the size of such objects is sufficiently large that one can locate and "handle" atoms and molecules with a high degree of certainty. The strongest support for this view is that by 1990 scientists had accumulated abundant evidence that atoms and molecules are routinely moved about with a high degree of accuracy by living systems and that researchers themselves had already begun to accomplish similar deeds in the laboratory.

Molecular Motion

A second concern about molecular nanotechnology is based on the fact that atoms and molecules are in constant motion. They may travel through space, vibrate back and forth, rotate, and move in other ways. This motion is sometimes referred to as "kT" movement because it can be expressed as a product of Boltzmann's constant, k, and the temperature, T, of the object in question.

Some early critics of molecular nanotechnology argued that the natural motion of atoms and molecules is sufficiently great that structures built by means of nanotechnological processes would not remain stable. They would shake and vibrate so much that they would fall apart during construction.

It can be shown that the amount of vibration that takes place in the atoms and molecules that make up a solid (although not a liquid and certainly not a gas) is small enough to cause no serious problems in the manipulation of such particles. Again, the best possible evidence for that position can be found in the molecules that make up living cells, such as nucleic acid and protein molecules. These molecules are routinely created and moved about in cells to build new structures, just as has been suggested for molecular nanotechnology. Yet the error rate in biochemical reactions is astonishingly low, of the order of one part in 10^{11} in some reactions that have been studied.

Early responses from the scientific community to *Engines of Creation* convinced Drexler that it would be necessary for him to prepare a more detailed and more technical exposition of his ideas. He had to write a book in which the specific physical and chemical principles underlying molecular nanotechnology were laid out in a mathematically sound manner. He also had to acknowledge and respond to technical questions about molecular nanotechnology, such as uncertainty and thermal motion, that his critics had raised.

The result was a book entitled *Nanosystems: Molecular Machinery, Manufacturing, and Computation*. *Nanosystems* differed dramatically from *Engines of Creation* in that its intended audience was not the general public, but the scientific community. The book is filled with mathematical proofs for the possibility of assemblers, replicators, nanocomputers, and other nanoscale devices. A layperson without a sound background in physics and chemistry quickly becomes hopelessly lost in the early pages of *Nanosystems*. Drexler wrote *Nanosystems* not only to buttress his arguments for molecular nanotechnology, but also to provide specific and concrete arguments on which his colleagues could evaluate and critique his ideas.

Drexler has always argued that his ideas about molecular nanotechnology—like any other scientific idea—can only benefit from the thoughtful analysis and criticism of his peers. Many of the attacks on *Engines of Creation* were and have been based on *ad hominem* arguments that "it will never fly" or "it just can't happen." Most new technologies have been met with attacks of this kind, attacks that contribute little or nothing to a real assignment of the new technology's potential. Molecular nanotechnology could, of course, turn out to be an unrealizable hope. But discovering that truth will come only from the tried-and-true process of experimental test, not from out-of-hand rejection of basic hypotheses.

Nanosystems was awarded the 1992 Association of American Publishers award for the year's best computer book. That award did not save the book from many of the same kinds of attacks leveled at *Engines of Creation*, however. For example, British chemist Fraser Stoddart was quoted as saying that Drexler's book was "a big disappointment for me! In it, he reveals that his knowledge of chemistry is scant to say the least" (as quoted in Regis, 1995, 265). Stoddart, interestingly enough, was one of the first researchers to have shown the possibility of manipulating individual atoms and molecules (Pollack, 1991, C1, 1).

Criticism also came from Roald Hoffman, of Cornell University, widely regarded as one of the brightest and most prescient researchers in chemistry. Drexler's molecular machinery, he observed, was "dull. It's the same as bigger structures, just tiny" (as quoted in Regis, 1995, 265). To Kurt Mislow, Professor of Chemistry at Princeton University, Drexler's ideas "far from being prophetic," were actually "retrogressive. These machines of his are all mechanical devices; they're crude, almost medieval, I would have to say—hopelessly medieval machinery" (as quoted in Regis, 1995, 265–6).

Nanosystems did meet the objections of some critics. They were able to see the precise scientific and engineering bases on which Drexler's vision of molecular nanotechnology was based. The book did not answer all conceivable questions about molecular nanotechnology, but it did show that a scientific argument could be made for most of the fundamental ideas originally proposed by Drexler.

Reasons for Optimism

Arguably the most important point buttressing Drexler's vision of molecular nanotechnology is the supporting data that self-replicating molecular machines already exist in abundance. They are found in all living organisms and are responsible for most changes that occur in those organisms.

Two of the most important and best understood classes of biomolecular machines are the nucleic acids (DNA and RNA) and proteins. By studying the structure and function of these two classes of biochemical molecules, one can better understand how human-made molecular machines might be constructed and made to function.

DNA

DNA is a large molecule consisting of a backbone of alternating units of the five-carbon sugar deoxyribose and a phosphate group. Attached to this backbone is a series of nitrogen bases (adenine [A], cytosine [C], guanine [G], and thymine [T]), whose specific sequence can be thought of as an instructional code for the cell. The sequence -A-G-G-C-C-G-G-C-C-C-A-C-C-, for example, is read by a cell to mean: "make the amino acid serine," "then make the amino acid glycine," "then make the amino acid arginine," and so on. In this regard, DNA can be thought of as a primitive nanocomputer—primitive because it contains only a limited message directing chemical changes in a cell.

DNA also can be thought of as a replicator. At certain points in the life of a cell, a DNA molecule divides into two parts along the long axis of the chain. Each of the two parts formed by this division then makes an exact copy of itself. Two new molecules are formed that are identical to the original parent molecule.

The instructions carried in the DNA nanocomputer are used to build new and different molecules of RNA. In this regard, DNA molecules are also assemblers that take raw materials in the cytoplasm of a cell and put them together to form a new molecule of RNA. The RNA thus formed then repeats this process by looking for certain raw material molecules in the cell and assembling them to make a protein molecule. RNA molecules are assemblers also.

Proteins

Proteins are another example of molecular machines found in all living cells. Proteins are large molecules consisting of tens of thousands of molecules, or more. They are formed from long strands of amino acids (as directed by the message stored in DNA molecules) that spontaneously fold into distinctive three-dimensional structures. The molecule in Figure 1.2 illustrates a typical three-dimensional conformation found in many protein molecules.

Protein molecules perform a host of essential functions in living organisms. They are used as the building blocks from which cells are made, they transport essential atoms and molecules from one part of the organism to another, and they make possible most of the chemical reactions (such as energy production) on which life depends. The specific function carried out by a protein molecule depends on and is made possible by the protein's structure, just as the function of any

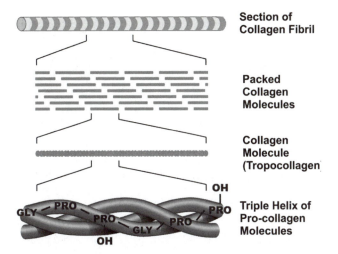

Figure 1.2. The triple-helix configuration of collagen molecules. *Courtesy of Charles Currey, University of Florida.*

human-made molecular machine is determined by its molecular composition and conformation.

Enzymes are among the best examples of the way in which nature uses molecular machines (proteins) to carry out essential functions (the assembling of new molecules). An enzyme is a protein that acts as a catalyst (a substance that speeds up chemical reactions) in a living organism. It accelerates the rate at which:

- Two molecules are joined to each other to make a new molecule.
- One molecule is broken apart into two new molecules.
- An atom or group of atoms is added to or removed from a molecule.
- Some other modification is made in the structure of a molecule.

There are a variety of mechanisms by which enzyme molecules bring about these changes. In general, an enzyme molecule attaches itself to the molecule (or molecules) on which it acts (the *substrate*). In some cases, the substrate fits neatly into a cavity in the enzyme molecule. In other cases, the enzyme molecule may undergo changes in its conformation to accommodate the substrate molecule.

Chemical bonds eventually form between the enzyme molecule and the substrate molecule(s). At this point, a variety of changes may occur to effect a change in the substrate molecule. For example, bonding between the enzyme and substrate molecules may weaken other chemical bonds within the substrate molecule, causing it to break apart.

The important point is that the enzyme molecule acts as an assembler or a disassembler. It takes one or more molecules and builds a new molecular structure from it or them. For those interested in molecular nanotechnology, enzyme molecules provide an attractive model as to how a Drexlerian assembler might be constructed and how it might work.

Other Biomolecular Machines

Scientists are now learning a great deal about the structure and function of nucleic acids and proteins. One of their most remarkable successes has come with an enzyme known as (*ATP*) *synthase*, or *ATPase*. ATP synthase is an enzyme that catalyzes the production of the compound adenosine triphosphate (ATP) in cells. ATP is an important biochemical molecule that provides the energy needed to drive many biochemical reactions. It is produced in cells when ATP synthase adds a single phosphate group (P_1) first to a molecule of adenosine monophosphate (AMP) to form adenosine diphosphate (ADP), then repeats the process to form ATP:

$$AMP + P_1 \rightarrow ADP$$
$$ADP + P_1 \rightarrow ATP$$

During the 1950s, American chemist Paul Boyer hypothesized a method by which this reaction might occur. Boyer's theory was confirmed experimentally in the 1990s by British chemist John E. Walker. According to the Boyer/Walker model, a molecule of ATP synthase consists of two parts, called F_0 and F_1. The F_0 segment is attached to a cell membrane, and the F_1 segment protrudes outward from the F_0 segment. The F_1 segment, in turn, consists of two parts, a "cylinder" and a "rod" running through the center of the cylinder.

When the rod rotates inside the cylinder, it picks up a phosphate unit in the cytoplasm of the cell and attaches it to an AMP or an ADP molecule. Final confirmation of this remarkable system was obtained in 1997 when a Japanese team of researchers photographed the rotation of the rod unit within the cylinder of the ATP synthase molecule. They referred to the structure as "the smallest biological rotary motor known." In recognition of this remarkable research, Boyer and Walker were jointly awarded the 1997 Nobel Prize in Chemistry.

The point of the above-mentioned examples is that machine-like structures consisting of only a few hundred or thousand atoms already exist within living cells. These structures function efficiently without disruption because of quantum or vibratory effects. When errors occur or molecules "break down," systems exist for repairing the damaged molecules. The fact that such systems do exist and do function well suggests that artificial systems, such as assemblers and replicators, should also be possible.

Research in Molecular Nanotechnology

The discipline of molecular nanotechnology is different from other fields of science in that it consists of two different approaches to the subject of molecular devices. One group of researchers is interested primarily in a long-term vision as to what a mature science of molecular nanotechnology might look like. That view is similar to the one expressed by Drexler in *Engines of Creation*. It focuses on the design of assemblers, replicators, nanocomputers, and their components and the methods by which such devices might be made.

The primary constraint on this line of research is that there are virtually no concrete objects or data (other than those found in living organisms) with which to carry out experiments. One cannot design an experiment in which an assembler, a replicator, or a nanocomputer is subjected to various kinds of tests. The only avenues of research available to this first category of researchers are molecular modeling and theoretical calculations.

Fortunately, scientists already know a great deal about the properties of atoms and molecules. They know the physical dimensions of such particles, their tendencies to react or not react with each other, the energy needed to make reactions occur, the physical conformations that particles may take at rest and during reactions, and so on. This information can be translated into mathematical equations and models and into computer programs that allow visual images of atoms and molecules to be manipulated in realistic ways. Researchers can build ideal, computer-generated molecular machines; study their properties; and find out how they might be expected to operate in the real world. Some of the molecular devices that have been designed for testing and study are shown in Figure 1.3. Scientists engaged in this line of research are not focusing on molecular structures that may be built and tested in the next year or two, but rather on the kinds of devices that may not be available for another decade or more.

Figure 1.3. Computer model of a molecular differential gear. © 1997 *Institute for Molecular Manufacturing (IMM)* <*www.imm.org*>. *All rights reserved.*

A second and much larger group of researchers is interested in experiments and inventions that can be designed and carried out now or in the near future (no more than a year or so from now). These researchers work with the tools now available for the construction and manipulation of nanosize devices, or they invent new tools or modify existing devices to use in molecular nanotechnology studies. They also focus on existing structures and systems that might someday form the basis of assemblers, replicators, nanocomputers, and other molecular devices. Some of the approaches followed by those involved in this line of research include:

- Examining ways of modifying natural DNA and proteins or constructing artificial analogues of such molecules to make them capable of carrying out various machine-like functions.

- Studying the electrical properties of atoms and molecules to see how such particles can be used in electrical circuits.

- Learning more about the way naturally occurring molecules self-replicate or correct errors that occur during assembly or self-replication.

- Finding out how the properties and behaviors of nanosize particles differ from those of larger size and how those properties and behaviors can be used in the design and construction of molecular machines.

As in all fields of scientific research, the theoretical and experimental are by no means distinct from each other. Theory suggests lines of experimentation, and new information from experiments suggests modifications in theory. This interconnection in molecular nanotechnology was recognized in the 1999 report of President Clinton's National Nanotechnology Initiative:

Experimentation and modeling/simulation capabilities will be equally important to advances in understanding, each testing and stimulating the other, compelling the development of new computational methods, algorithms and high performance computing resources. Modeling and simulation at nanoscale will enable new synthesis and processing methods of nanostructures, control of nano-manipulators such as atomic force microscopes, development of scale-up techniques, and creation of complex systems and architecture based on nanostructures. (*National Nanotechnology Initiative*, 2000, 75)

A few decades ago, the challenges faced by both lines of research in molecular nanotechnology were enormous. Theorists had a relatively good understanding of the properties of atoms and molecules, but little by way of computer modeling programs with which to construct devices based on this knowledge. Experimentalists had virtually no way of manipulating matter at the atomic or molecular level. The task of building anything atom-by-atom—theoretically possible or not—was simply out of the realm of possibility.

In the last decade, advances in both theoretical and experimental molecular nanotechnology research have been enormous. Today, there are many reliable computer programs that allow researchers to construct and manipulate molecules and molecular structures with a large range of sizes and shapes. Researchers can find out precisely how nanodevices might be built, what they will look like, and how they will operate. Theorists can easily make adjustments and modifications in their models and see how such changes will affect the properties and behavior of a device.

Experimentalists now routinely manipulate atoms and molecules with a precision thought to be impossible only twenty years ago when *Engines of Creation* was written. They have constructed and studied single-molecule logic systems, wires no more than a few nanometers in diameter that can carry an electric current, and artificial nucleic acid and protein molecules capable of carrying out a whole range of new functions.

In summary, there are still many highly regarded scientists who consider Drexler's vision of assemblers, replicators, and nanocomputers as an unattainable fantasy, an intriguing tale of science fiction, but an unrealistic view of humankind's future. The fact remains, however, that the dawn of the 21st century has seen rapid progress in the development of many techniques, materials, and devices on which a Drexlerian vision of molecular nanotechnology would ultimately be based. Chapter 2 provides a more detailed review of the types of research now being conducted in the field.

References

Amato, I. *Nanotechnology: Shaping the World Atom by Atom.* Washington, DC: National Science and Technology Council, September 1999.

Byrne, Peter. "Small Wonders." *SF Weekly* (8–14 December 1999): 15–23.

Drexler, K. Eric. "Molecular Engineering: An Approach to the Development of General Capabilities for Molecular Manipulation." *Proceedings of the National Academy of Sciences* (September 1981): 5275–8.

Drexler, K. Eric. *Engines of Creation: The Coming Era of Nanotechnology.* New York: Anchor Books, 1986. The book is also available in its entirety on the Internet at http://www.foresight.org/EOC/.

Drexler, K. Eric. *Nanosystems: Molecular Machinery, Manufacturing, and Computation.* New York: John Wiley & Sons, 1992.

Feynman, Richard P. "There's Plenty of Room at the Bottom." *Engineering and Science* (February 1960): 22–4.

"Foresight Debate with *Scientific American*." Available at http://www.foresight.org/SciAmDebate/SciAmOverview.html.

Garfinkel, Simon. "Critique of Nanotechnology." *Whole Earth Review* (Summer 1990): 104–13.

Jones, David E. H. "Technical Boundless Optimism." *Nature* (27 April 1995): 835–7.

Kuhn, Thomas. *The Structure of Scientific Revolutions,* Second Edition. Chicago: University of Chicago Press, 1970.

National Science and Technology Council, Subcommittee on Nanoscale Science, Engineering, and Technology. *National Nanotechnology Initiative.* Washington, DC: [n.p.], 2000.

Pollack, Andrew. "Atom by Atom, Scientists Build 'Invisible' Machines of the Future," *New York Times* (26 November 1991): C1, 1.

Regis, Ed. *Nano: The Emerging Science of Nanotechnology.* Boston: Little, Brown and Company, 1995.

Stix, Gary. "Waiting for Breakthroughs." *Scientific American* (April 1996): 94–9.

Chapter Two

The Promise and Threat of Nanotechnology

One can scarcely read an article about molecular nanotechnology—whether in the scientific or popular press—without finding some discussion of potential benefits to be expected from developments in the field. Those benefits range from the relatively mundane and near-at-hand to those that may not be available for decades and that are far more revolutionary in character.

Articles written for the general public often emphasize the more dramatic, long-term products of research in molecular nanotechnology. They often draw on Drexler's *Engines of Creation* for their vision of a molecular technology–enhanced future. In that book, Drexler outlined some of the ways in which this new field of science and technology might create a golden future for the human race.

The ultimate product of molecular nanotechnology most frequently cited by writers is probably the "breadbox assembler." The breadbox assembler is a relatively small (breadbox size) device containing thousands of assemblers, replicators, and nanocomputers. The nanocomputers carry instructions for the manufacture of a host of common household products, such as pots and pans, hairbrushes and mirrors, gloves and underwear, and tennis rackets and baseballs. About the only limitation on the kinds of objects for which the nanocomputer can be programmed is the size of the box in which they are made. To make larger objects, such as a bookcase, one would need to build a larger breadbox assembler.

To operate the breadbox assembler, a supply of raw materials is introduced into the box. The raw materials consist of common, readily available, inexpensive substances, such as sand and coal dust. The operator selects the object to be manufactured from a list of those that can be produced. Molecular devices inside the breadbox assembler then begin their work of assembling the desired product by following the instructions stored in the nanocomputer. After some period of time, assembly is complete, and the final product is ejected from the machine. The similarity between this process and the manufacture of proteins, albeit on a dramatically different scale, is obvious.

The availability of a breadbox assembler in every person's home would have profound, perhaps unimaginable, consequences for human civilization. For the cost of the breadbox assembler itself and the raw materials needed for its operation, any individual or family could manufacture most of the objects they would need for a safe, healthy, reasonably prosperous life. The most basic human needs—food, for example—could be met by anyone in the world at the minimal cost. Under such circumstances, the state of poverty might disappear or be dramatically altered.

Space travel is another area that would benefit from a mature molecular nanotechnology. The greatest challenge to human space travel today is the cost of providing large spacecraft with enough energy to overcome the pull of Earth's gravity. Molecular nanotechnology provides a way of building spacecraft and other devices used in space travel from materials that are many times stronger and yet many times lighter than anything now available—or anything that probably can ever be built with existing "top-down" technology.

The products of molecular nanotechnology can be employed in almost every area of space travel. In *Engines of Creation*, Drexler described a "smart" spacesuit that is little more than a "second skin." It is only a few millimeters thick with an inner layer consisting of molecular devices that can perform a variety of functions as directed by localized nanocomputers. In Drexler's description:

> The space suit of the future has the strength of steel and the flexibility of your own body. If you reset the suit's controls, the suit continues to match your motions, but with a difference. Instead of simply transmitting the forces you exert, it amplifies them by a factor of ten. Likewise, when something brushes against you, the suit now transmits only a tenth of the force to the inside. You are now ready for a wrestling match with a gorilla.

What is more, the suit is durable. It can tolerate the failure of numerous nanomachines because it has so many others to take over the
load. The space between active fibers leaves room enough for assemblers and disassemblers to move about and repair damaged devices.
The suit repairs itself as fast as it wears out. (Drexler, 1986, 91, 92)

The first long-term Drexlerian prediction to become reality may involve "engines of healing," the title of Chapter 7 of *Engines of Creation*. Engines of healing are nanodevices consisting essentially of two
parts. One part is a nanocomputer programmed to carry out one or a
few specific functions such as instructions: "Look for abnormal DNA"
or "Look for cancer cells" or "Look for damaged blood vessels."
    The second part of the device contains molecular machines capable
of repairing cellular and molecular damage. Once the device (sometimes called a *nanorobot* or *nanobot*) has located its target (DNA, cell,
or tissue) in the body, repair and reconstruction can begin. Molecular
and cellular damage can be detected by such nanodevices long before
they can be found by any existing technology, such as x-ray, nuclear
medicine diagnostics, or blood tests.
    The potentials for nanomedicine are staggering. It should be possi-
ble not only to detect and repair molecules and cellular damage that
leads to any disease or disorder, but also to detect and deter the signs
of aging itself. The latter promise has caught the attention of those
interested in life extension, the search for methods that will allow humans to live well beyond the Biblical three-score-and-ten—perhaps
even to achieve immortality. (For more detailed information on nanomedicine, see the first book on the subject by Robert A. Freitas, Jr.,
Nanomedicine: Volume I: Basic Capabilities, available on the Internet
at http://www.nanomedicine.com.)

Short-Term Benefits

For the most part, the products of molecular nanotechnology described in *Engines of Creation* are not expected for many years or even
many decades. That fact makes it more difficult for many people to as-
similate, understand, and accept the possibility of such products. They
become easy targets for the critics of molecular nanotechnology.
    Not every product of molecular nanotechnology is 50 years dis-
tant, however. Experimentalists in the field are already envisioning
and talking about benefits from their research that might see the light
of day within a matter of years.

Perhaps the best collection of such predictions is the report on the National Nanotechnology Initiative (NNI), issued in July 2000 by the Subcommittee on Nanoscale Science, Engineering, and Technology of the National Science and Technology Council. In a review of the NNI's Grand Challenge grants, the report's authors list dozens of near-term benefits of research in molecular nanotechnology, as follows:

- Paints that change color as a function of temperature to provide automatic heating and cooling effects in homes.
- Automobile, aircraft, and spacecraft materials that incorporate sensors and repair devices that automatically detect and correct weaknesses that could lead to material damage and failure.
- "Stealth" materials that can detect radar and sonar signals and that can modify the materials themselves in such a way as to prevent the reflection of such signals and detection of an object.
- "Smart" surfaces that will last longer than any existing material; that are self-cleaning; and that can provide camouflage in the visible, infrared, ultraviolet, and other spectral regions.
- Biological nanodevices that can detect tumors that have affected not more than a few cells or have only moderately affected blood flow.
- Biological nanodevices that can detect DNA structures that may determine a person's susceptibility to certain diseases, infections, or toxins.
- Implantable nanodevices that are capable of repairing or assisting organs or organ parts and that are compatible with the body's biological and chemical environment.
- Drugs and other pharmacoactive substances that are normally insoluble in aqueous (water) solutions but that can be delivered directly to their desired sites of action.
- Nanoparticles that can be delivered to the DNA of cells for use in gene therapy.
- Nanomaterials capable of removing the smallest contaminants from water (<200 nm in diameter) and air (<20 nm in diameter).
- Efficient photovoltaic cells made of nanocrystalline materials that can be tuned to a wide range of the visual spectrum.
- Nanocrystals for use as highly efficient catalysts in the degradation of contaminants in polluted air and water.

- Networks of carbon nanotubes for use in hydrogen storage devices needed in hydrogen-based energy systems.

- Materials made of nanoparticles containing aluminum and aluminum oxide that are more resistant and lighter than the best steel alloy currently available.

- Nanoscale materials with magnetic properties four times better than those of any currently available permanent magnet.

- Materials made of nanoparticles capable of self-repair, self-adjustment to thermal properties, and active control of surfaces, as required in a variety of space applications.

- Bionanodevices that automatically detect and provide alerts to minor changes in body chemistry leading to mental and physical deterioration that tend to occur in a variety of stressful situations, such as flying in an airplane, long-distance driving, military actions, and police operations.

- Nanosensors implanted in building materials that monitor the status of such materials, detect developing flaws, report on fatigues and other damage, and, perhaps, initiate their own repair.

- Advanced nanobased military computer systems that make possible instantaneous worldwide communication, threat identification, secure encryption, speech recognition and language translation, and combat identification.

Potential Threats of Molecular Nanotechnology

One can easily become rhapsodic about the many potential benefits of advances in molecular nanotechnology. Like almost any field of science, however, molecular nanotechnology is a double-edged sword. It holds at least as much potential for the disruption of human civilization as it does for its advancement.

Drexler has thought and written about the potential risks of molecular nanotechnology for more than 20 years. In fact, one might argue that his concerns about the social implications of molecular nanotechnology have been at least as important as his interest in research in the field. In his "Afterword" to the 1990 edition of *Engines of Creation*, Drexler wrote that "Some have mistakenly imagined that my aim is to promote nanotechnology; it is, instead, to promote understanding of nanotechnology *and its consequences*, which is another matter entirely" (Drexler, 1986, 241). When Drexler and his colleagues founded the Foresight Institute in 1986, their goal was not primarily to promote research in molecular nanotechnology but rather to advance discussion of the ways in which this technology would affect human life and what could be done to deal with potential threats it posed.

Drexler's first extended discussion of the threats posed by molecular nanotechnology was in Chapter 11 of *Engines of Creation*, "Engines of Destruction." The tools and methods of molecular nanotechnology, he pointed out in that chapter, will clearly be as available to national governments as they will be to corporations and individuals. History has made it abundantly clear how national governments can put new technologies to use in the development and strengthening of their military systems.

One use of molecular nanotechnology would be in the rapid, automated, and relatively inexpensive production of currently available weapon systems. Assemblers, replicators, nanocomputers, and other molecular machines could be put to work in churning out vast amounts of weaponry as easily as the more benign consumer products. This option would not be limited to the strongest, richest, and most powerful nations. Like the breadbox assembler, the large-scale machinery of molecular nanotechnology would be simple and inexpensive enough for all but the poorest nations to afford. Just as the nuclear powers of the 20th century worried about the proliferation of nuclear technology to Third World (and even most First World) nations, so will *all* nations of the 21st century have to be concerned about the spread of molecular nanotechnology throughout the world.

The development of molecular nanotechnology also expands the possibilities for almost any government beyond the mere production of conventional weaponry. New kinds of weapons, similar to chemical and biological weapons—what Drexler called "programmable germs and other nasty novelties"—will also become possible (Drexler, 1986, 174). With a modest expenditure and relatively simple infrastructure, virtually any government should be able to manufacture such "germs" and "novelties" to an extent that would pose a significant threat to the most powerful nations of the world. The threat posed to the United States today by small terrorist groups operating from remote locations in the world's poorest nations provides a hint as to the kinds of dangers that a mature molecular nanotechnology might pose in the future.

Drexler also pointed out the dangers that molecular nanotechnology might pose for citizens of an oppressive government. We need hardly be reminded that some governments in the past have treated at least some of their own citizens as brutally as they have those of other nations. Efforts by the National Socialist (Nazi) government in Germany in the 1930s and early 1940s to wipe out Jews, gypsies, homosexuals, the disabled, and other "undesirables" in their own nation is perhaps only the best-known recent example of this phenomenon. The genocidal policies of Slobodan Milosevic and his colleagues against the Kosovar Muslims in their own country is another example. The availability of molecular nanotechnology can make monstrous efforts such as these even more efficient than they have been in the past.

Even if genocide is not the objective, oppressive governments will find molecular nanotechnology a powerful weapon of social control. It should eventually be a simple matter to disperse nanobots into the atmosphere with the ability to monitor the location, conversation, and behaviors of any or all citizens. The same nanobots used to sense and repair cell damage will also be able, according to Drexler, to "tranquilize, lobotomize, or otherwise modify entire populations." To those who question whether such extreme possibilities at all exist, Drexler noted that "the world already holds governments that spy, torture, and drug; advanced technology will merely extend the possibilities" (Drexler, 1986, 176).

Critics may point out that the preceding scenario—similar to much of what Drexler writes about molecular nanotechnology—is wildly implausible. That may or may not be true. Drexler's position is that discussion about the benefits and the risks of a mature molecular nanotechnology should be started now, well before that technology is widely available. We could, perhaps, accept the proposition that Drexler's views are entirely absurd and that humans will never have to deal with the issues posed by medical nanobots, breadbox assemblers, and programmable germs. If that should be the case, no harm will have been done. If it turns out not to be the case and Drexler is at least partially right in his predictions and concerns, human civilization will have little time to think about and decide how to deal with what might become the most revolutionary technology ever developed by the human species.

References

Drexler, K. Eric. *Engines of Creation: The Coming Era of Nanotechnology*. New York: Anchor Books, 1986. The book is also available in its entirety on the Internet at http://www.foresight.org/EOC/.

Freitas, Robert A., Jr. *Nanomedicine: Basic Capabilities.* Austin, TX: Landes Bioscience, 1999.

National Science and Technology Council, Subcommittee on Nanoscale Science, Engineering, and Technology. *National Nanotechnology Initiative.* Washington, D.C.: [n.p.], 2000.

Chapter Three

Recent Advances

What kind of research will lead to the first assembler, the first replicator, the first nanocomputer? No one can give a definitive answer to that question. As with many lines of scientific research, it does not seem possible to design a specific series of experiments that will result in the first fully functioning molecular device. Even if such a plan of research were possible, it seems unlikely that researchers themselves all would agree to follow some agenda developed and imposed by some external organization or agency.

In molecular nanotechnology research, groups of scientists are working on specific problems of greatest interest to themselves, sometimes producing results that fit into the long-range evolution of molecular nanotechnology. In many cases, these researchers may not be especially interested in larger problems, such as the design and construction of an assembler, a replicator, or a nanocomputer. They may reject the notion that such devices can even be built. Some of the most severe critics of a Drexlerian view of molecular nanotechnology are researchers whose own work would seem to be important steps toward the creation of nanomachines.

This chapter provides a review of more recent research that reasonably might be expected to lead to the development of nanoscale devices, components of which assemblers, replicators, and nanocomputers eventually might be constructed. The chapter is divided into five main sections. The first section deals with arguably the most important tool available to molecular nanotechnologists—the scanning probe microscope (SPM). *SPMs* refer to a group of instruments that allow one to observe and manipulate individual atoms and molecules.

The second section reviews progress in the field of nucleic acid and protein studies, naturally occurring molecular machines whose structures and functions are widely believed to be a clue to the design and operation of artificial molecular machines. The third section summarizes some of the research on carbon nanotubes, structures thought to hold potential as one of the major building blocks of nanoscale devices.

The fourth section discusses progress in the field of molecular electronics, a field whose primary goal is to find ways of building electronic computers with dimensions many times smaller and operating speeds many times greater than the best computing devices currently available. The final section is a summary of research on other kinds of nanoscale devices, including nanosize pens, motors, wheels, tweezers, bearings, and springs.

Basic Tools: Scanning Probe Microscopes

Perhaps the most daunting challenge in molecular nanotechnology has been to find a way of manipulating individual atoms and molecules. If one assumes that bottom-up nanotechnology is possible, the first question that arises is how one will be able to pick up atoms and molecules, move them from place to place on a surface, and put them down in some precise location.

One solution to that problem appeared unexpectedly in 1981 with the invention of the scanning tunneling microscope (STM) by Heinrich Rohrer and Gerd Karl Binnig, researchers at IBM's Zurich Research Laboratory (IBM-ZRL). Rohrer and Binnig were awarded the 1986 Nobel Prize in Physics for their work.

The STM was invented as an observational tool, a device for examining the structure of materials at magnifications higher than those possible with an optical (light) microscope. The design of an STM in principle is relatively simple. The working part of the microscope consists of a small tip, only a few atoms or molecules in diameter, attached to a piezoelectric shaft. A piezoelectric material is one whose dimensions change when an electrical current flows through it.

To use the STM, an operator brings the tip of the instrument very close to the surface of the material being examined. "Very close" means a distance of about 1 nm or less. The tip normally does not come into contact with the surface because the force of separation between electrons in the tip and in the surface at distances of less than 1 nm is quite large.

At the tip's closest approach to the surface, electrons may flow from the instrument to the surface, or vice versa. According to the laws of classic physics, this flow of electrons is not possible because of the repulsion of like charges (in electron clouds composing the outer part of atoms) on the two materials. The laws of quantum mechanics do allow some electrons to "sneak through" or "tunnel under" that energy barrier, however, and an electrical current can be observed between surface and tip. The occurrence of this current gives the STM its name.

As tunneling current flows between microscope tip and surface at a specific location, its change can be detected and measured by the operator. The microscope is moved (scanned) across the surface of the material being studied by applying a current/voltage to the piezoelectric shaft, which changes in length. The topography of the surface is generated by plotting the location and the current.

One disadvantage of the STM is that it can be used only on conductive surfaces, such as those of metals. Many interesting materials, including most biological substances, are nonconductive, however. To overcome this problem, a modification of the STM known as the atomic force microscope (AFM) was invented in 1986 by Binnig, Christoph Gerber at IBM-ZRL, and Calvin Quate at Stanford University.

In an AFM, a cantilever and tip is attached to a microscope similar to an STM. The cantilever is a long, narrow, flexible strip of material, similar to a tiny diving board. As the AFM tip is passed over the surface of a material, the cantilever experiences weak forces of attraction or repulsion (van der Waals forces) and electronic repulsion between atoms, depending on its distance from the surface. These forces cause the cantilever to flex upward or downward, changes that can be detected and recorded by the microscope. Because these forces exist between all kinds of material, conductive or not, the AFM can be used to study a much wider range of objects than the STM.

The STM has been modified in many other ways. For example, some researchers have attached protein-receptor molecules to the tip of an STM or the end of the cantilever of an AFM. A protein-receptor molecule is a naturally occurring molecule that recognizes and bonds to a given protein molecule. When an STM or AFM modified in this way is passed over a surface containing the recognized protein, a bond forms between one protein molecule on the surface and the protein-receptor molecule on the tip or the cantilever. The operator can exert a force that plucks the protein molecule off the surface. One authority in the field has described this technique as being similar to fly-fishing (Weiss, 1998, p. 269).

Another modification of the STM and AFM was developed to deal with one of the basic problems in electron microscopy: chemical identification of atoms and molecules present on a surface. A basic STM can provide a good map of the topography of a surface, but it is not able to identify easily the chemical species that make up that map. Modest success of differentiation of atoms on the surface was achieved by varying the bias voltage between the tip and the surface.

One method for solving that problem was developed by Wilson Ho and colleagues at Cornell University in 1998. Ho's approach was based on the fact that atoms and molecules sometimes can be identified by the nature of their chemical bonds, specifically, by the vibrational energy of those bonds. Ho hypothesized that the atoms and molecules on a surface might be identifiable if the tip of an STM could be brought close enough to them to detect their vibrational energy and if other disturbing effects (such as the kinetic motion of the particles themselves) could be eliminated or controlled. In their first experiments, Ho's team studied acetylene molecules deposited on a smooth copper surface with an STM whose tip was brought to less than one thousandth of a nanometer from the surface. At this distance, the vibrational energy of the carbon-hydrogen bonds in acetylene was detectable and was consistent with that known for the molecule.

Other modifications of the STM and AFM have been invented to improve the identity of chemical species; to study the nature of chemical reactions that occur between molecules and atoms; and to measure the elasticity, humidity, magnetic and frictional properties, and other physical characteristics of materials. Today the variety of STM and AFM modifications often is known by the general designation of *scanning probe microscopes*.

For those interested in molecular nanotechnology, the most exciting developments in SPMs has been the design of instruments capable of manipulating individual atoms and molecules. Rather than functioning entirely as passive devices for observing and reporting on the structure of materials, some SPMs have become active instruments capable of moving, lifting, carrying, pushing, and positioning individual atoms and molecules.

The first major breakthrough in this area occurred in 1989, when Donald M. Eigler and E. K. Schweizer at IBM's Almaden Research

Center in San Jose, California, positioned 35 atoms of xenon on a nickel surface to spell out the company's logo, "IBM." These researchers cooled their apparatus to a temperature of about 3 K (−270°C) to reduce thermal vibration of the atoms and to make them stable enough to manipulate. They then used an STM in its imaging mode to search for xenon atoms on the nickel surface. When an atom was located, the STM was lowered until the tunneling current reached a maximum, at which point the xenon atom was attracted more strongly to the tip than to the nickel surface. The tip was then moved across the nickel surface, dragging the xenon atom into its desired position. Finally, the STM was switched back to its imaging mode, the force between xenon atom and tip was greatly reduced, and the xenon atom was fixed to its new position. As simple as the process may sound in this description, the actual manipulation of xenon atoms was time-consuming. At the time, it seemed difficult to imagine how the process could be adapted to the manufacture of any kind of useful device.

The IBM experiment illustrated one fundamental challenge in manipulating atoms and molecules: thermal motion. In many cases, atoms and molecules are vibrating, rotating, or moving in other ways too rapidly to allow their capture or control by an SPM. The IBM solution—cooling xenon and nickel to nearly absolute zero—solves this problem. The solution is not satisfactory, however, for the widespread manufacture of assemblers, replicators, nanocomputers, and other nanodevices, work that must be done at or near room temperature.

That thermal motion is not an insurmountable problem to the manipulation of atoms and molecules was shown first by another IBM team in 1996 (and many other research groups since). The IBM team, from IBM-ZRL, in conjunction with researchers from the French National Center for Scientific Research (CNRS), began its work with a search for a molecule with certain specific properties. First, the molecule had to bond tightly enough to a surface for it to stay in position. Second, the bond between molecule and surface had to be weak enough to allow an SPM to pick up and move the molecule without tearing the molecule apart. Third, and as a corollary, the bonds within the molecule had to be strong enough to prevent the molecule from breaking apart as it was being manipulated.

After a study of many candidate molecules, the IBM/CNRS team selected a compound known as copper-tetra-(3,5 di-tertiary-butyl-phenyl)-porphyrin (Cu-TBP-porphyrin). Cu-TBP-porphyrin consists of two parts, the core of which is a porphyrin ring. The porphyrin ring occurs in many naturally occurring molecules, including the blood-carrying compound, hemoglobin, and chlorophyll, the compound that catalyzes the process of photosynthesis. In the Cu-TBP-porphyrin molecule, four identical hydrocarbon chains are attached to opposite corners of the ring, forming a table-like structure. When the molecule is nudged with the tip on an SPM, it appears to wobble across the surface, a bit like a four-legged ant with arthritis. As with much SPM work, the IBM/CNRS team used its SPM in imaging and manipulation modes alternatively, allowing it to move the molecule, then, by changing the distance between tip and service, observing the change that had occurred.

The series of six SPM images in Figure 3.1 show the changes that occurred in this experiment. Note the random arrangement of Cu-TBP-porphyrin molecules in frame 1. In frames 2 through 5, the ordering effect of manipulation by the SPM gradually becomes apparent until, in frame 6, a coherent new structure consisting of six Cu-TBP-porphyrin molecules can be seen. This structure is entirely synthetic and does not occur in nature.

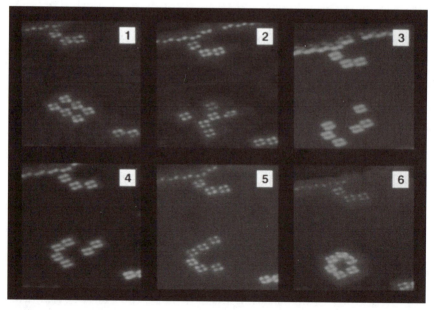

Figure 3.1.   These six panels show the changes that occurred during a 1996 IMB-ZBL experiment involving the manipulation of Cu-TBP-porphyrin molecules. *Courtesy of IBM Zurich Research Laboratory.*

By the end of the 20th century, the manipulation of atoms and molecules with SPMs had become routine. A team of researchers at Japan's Aono Atomcraft Project had reported on experiments in which they had used an STM to extract individual silicon atoms from the surface of a silicon crystal, move them to new locations, and bond them to the surface at the new location. The same atoms could be removed at a later time without damaging or dislocating the silicon surface. The researchers also reported success in repairing holes in the silicon surface with atoms brought from another location and in building simple structures composed of silicon atoms on the silicon surface.

At the same time, simple chemical experiments using SPM devices were becoming possible. In 1995, Tsung-Chung Shen and colleagues at the University of Illinois at Urbana-Champagne described an experiment in which individual chemical bonds were broken. The researchers began with a silicon crystal on which a monatomic layer of hydrogen was laid down. Bonds between silicon and hydrogen formed, yielding silicon monohydride (SiH) and silicon dihydride (SiH_2).

Researchers then moved the tip of an STM across the surface of the material, alternating voltages in the instrument so that it acted alternatively in imaging and manipulating modes. Under the latter condition, they were able to pluck individual atoms from SiH and SiH_2 molecules, leaving behind unbonded electron pairs ("dangling bonds") on silicon atoms. The presence of such bonds showed up as bright spots in the STM micrograph, prompting one researcher to point out, "We can actually get down to individual rows of molecules on a silicon surface and selectively remove hydrogen along these rows" (Lipkin, 1995, p. 391).

Another example of SPM-initiated chemical reactions was reported in 1999 by Wilson Ho and Hyojune Lee of Cornell. In this research, Ho and Lee deposited atoms of iron and molecules of carbon monoxide on a silver surface in a vacuum at a temperature of 13 K ($-260°C$). They used an STM to scan the surface and locate iron atoms and carbon monoxide molecules. When they encountered a carbon monoxide molecule, they lowered the tip of the STM and increased the voltage, causing a force of attraction between tip and molecule. At this point, Ho and Lee were able to pluck the carbon monoxide molecule from the silver surface and position it above an iron atom. When voltage in the STM tip was reduced, attractive forces between the iron atom and carbon monoxide molecule were strong enough for a bond to form between the two, and a new molecule of iron carbonyl [Fe(CO)] was formed. In a second stage of the research, Ho and Lee were able to repeat this process, adding a second molecule of carbon monoxide to an iron carbonyl molecule to form iron dicarbonyl [Fe(CO)$_2$].

Two decades after its invention, the STM and its cousins have become powerful tools in molecular nanotechnology research. They can be used as "nanocranes" capable of performing many of the basic functions needed in the manufacture of assemblers, replicators, and other nanodevices. Some of the many ways in which they currently are being employed in research are discussed later in this chapter.

Basic Materials: Nucleic Acids and Proteins

One response to critics of a Drexlerian view of molecular nanotechnology is that assemblers, replicators, nanocomputers, and other molecular devices already exist in living organisms and have functioned effectively for millions of years. They build new nanodevices from simple raw materials using complex instructions stored in molecule-size memory devices; they repair damaged nanosize devices in cells; they function efficiently despite quantum and kinetic effects and despite numerous errors that exist in molecular devices and the functions they perform; and they make accurate and precise copies of themselves (that is, they are replicators).

One potentially productive line of research in molecular nanotechnology might be to understand better the structure and function of existing nanomachines, such as nucleic acids and proteins, and to build modifications or synthetic analogues of these devices capable of performing one or more of the functions to be carried out by assemblers, replicators, nanocomputers, and other nanodevices. It is advantageous that many researchers with little or no interest in molecular nanotechnology are interested in these same problems and can contribute to this quest. For example, understanding protein structure and function is a crucial first step in the design of new drugs.

A review of the steps by which cells make proteins may illustrate the way this process is of interest in the field of molecular nanotechnology. The process begins with molecules of DNA, simple nanocomputers that carry information about the design of protein molecules. DNA molecules are large, spaghetti-like chains that consist of two strands intertwined with each other in a helical structure (see Figure 3.2). Each strand of the molecule is made of some extended

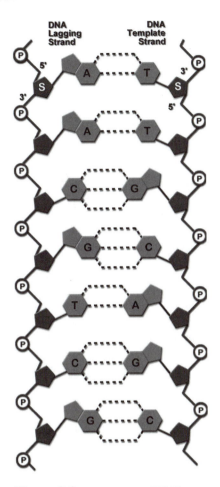

Figure 3.2. Structure of DNA molecule. © *Regents of New Mexico State University/Southwest Biotechnology and Informatics Center (SWBIC).*

sequence of nucleotides, structures consisting of a sugar (deoxyribose), a phosphate group, and one of four nitrogen bases. The four nitrogen bases found in DNA are adenine (A), cytosine (C), guanine (G), and thymine (T). The DNA molecule is held together by weak hydrogen bonds between nitrogen bases on the intertwined strands. The bonding follows a unique pattern: adenine and thymine bond only to each other, and cytosine and guanine bond only to each other.

The information stored in DNA molecules is coded by means of the sequence of nitrogen bases in the molecule. Each set of three nitrogen bases (a *codon*) codes for one of the naturally occurring amino acids (or, in a few cases, has no coding function). A nitrogen base sequence such as A–A–C–T–C–A–T–T–T–G–A–A, for example, is read by cells to mean: "make the amino acid leucine, then make the amino acid serine, then make the amino acid lysine, then make the amino acid leucine." Because DNA molecules are large with thousands of bases, and because those bases can be arranged in essentially any sequence, a large variety of coding instructions can be stored in such molecules.

DNA molecules also can be thought of as Drexlerian replicators in that they naturally and spontaneously make copies of themselves. In doing so, a DNA molecule "unzips," as shown in the upper portion of Figure 3.2. Each strand of the molecule then selects free molecules from the cell environment and arranges them in a sequence determined by the existing strand to make a new and precisely accurate complementary strand. As this process goes forward, two new DNA molecules are produced, both exact copies of the original DNA molecule. This process is precisely what one expects a Drexlerian replicator to do.

The process by which new proteins are made according to the instructions stored in DNA can be compared with the process by which cars are assembled in an automobile factory. In the first step of that process, the DNA molecule unzips, as shown in Figure 3.3. One strand of DNA is then used as a template to make another type of nucleic acid known as ribonucleic acid (RNA). This form of RNA is used to carry the protein-synthesis information stored in DNA to structures known as ribosomes in other parts of the cell, where proteins are made. Because of its function, this form of RNA is known as messenger RNA (mRNA).

mRNA is made when nucleotides in the cell environment are attracted to the exposed DNA template strand. Each "new" nucleotide bonds to a complementary nitrogen base on the DNA template following nitrogen-pair matching rules. For example, a guanine molecule from the cell environment bonds only to a cytosine base on the DNA template strand. Enzymes attach the newly assembled nucleotides to each other in a sequence dictated by the sequence of nitrogen bases on the DNA template. In this stage of the assembly line,

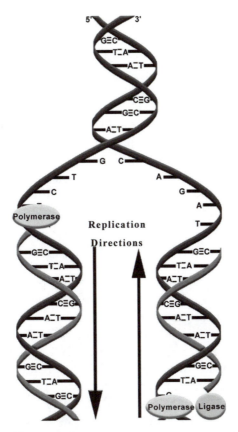

Figure 3.3. Transcription of DNA.
© *Regents of New Mexico State University/*
Southwest Biotechnology and Informatics
Center (SWBIC).

precise, but reversed, copies of the DNA code are stored in mRNA molecules, which then travel to the ribosomes.

The manufacture of new protein molecules requires four elements: (1) the coded instructions stored in DNA and transferred to ribosomes in mRNA molecules, (2) a supply of amino acids in the cell environment from which the new proteins can be assembled, (3) a set of "delivery trucks" to collect the amino acids and bring them to the construction site on the ribosome, and (4) an assembly line on which the amino acids can be put together in precisely the correct order as required to make a given protein.

The delivery trucks in this process are another form of RNA known as transfer RNA (tRNA). These molecules get their name from the fact that they transfer amino acids from the general cell environment to the ribosomes, where they become the raw material from which proteins are made. tRNA molecules are elongated molecules with a particular sequence of nitrogen bases at one end (e.g., C–A–C) and a "pocket" at the other end uniquely shaped to hold the amino acid for which the nitrogen base sequence codes (in this case, the amino acid histidine).

A tRNA molecule moves through a cell searching for an amino acid that will fit its particular pocket. It then carries that molecule back to the assembly line on the ribosome and looks for a section of an mRNA molecule waiting there that matches its (the tRNA's) nitrogen base sequence. Chemical bonds form between each tRNA and a complementary section of the mRNA. As a sequence of such complexes begins to develop, enzymes bind adjacent amino acids to each other, bonds between tRNA and mRNA and between tRNA and amino acids break, and a growing chain of amino acids—a protein—is released from the assembly site.

It takes little imagination to see how the process of protein synthesis might be used as a model for the development of assemblers, replicators, and other nanodevices. Current research in this field is aimed not only at an improved understanding of the structure and function of nucleic acids and proteins, but also at the modification of existing structures and the invention of new molecules not found in nature but capable of performing similar and enhanced functions.

Protein Research

Perhaps the most fundamental problem in protein research today is determining the structure of these molecules. Protein molecules are large, consisting of tens or hundreds of thousands of atoms. These atoms generally are arranged in long chains that twist and fold on themselves, sometimes in jumbled arrangements such as that of a telephone cord that has become tangled around itself. In many cases, protein molecules contain two or more subunits joined to each other by bridges of various kinds. The molecules may contain holes, tunnels, caves, and other characteristic structures. These structures are often responsible for the distinctive function of the protein molecule in question.

The problem for researchers is that "taking a picture" of a protein molecule is an extraordinarily difficult task. The most common procedure for studying protein structure is x-ray crystallography, a process in which a beam of x-rays is shined through a crystal of the protein being studied. The way in which the x-rays are deflected gives a clue as to the way atoms are arranged in the molecule. The presence of so many atoms in a protein makes their analysis by this method very difficult.

Still, impressive progress has been made in determining the structure of various proteins. A group of researchers at the University of California at Santa Cruz (UCSC) and the Lawrence Berkeley National Laboratory in Berkeley, California, reported in 1999 that they had determined the complete structure of a ribosome at a resolution of 7.8 A (0.78 nm). This research is important to the advancement of molecular nanotechnology because, as one reviewer noted, "[t]he ribosome is in some respects the closest thing to a general assembler that exists today" (Soreff, 1999, p. 10).

As noted previously, ribosomes are the cellular "factory" sites where proteins are assembled from amino acids. One of the UCSC researchers compared the ribosome to a "miniature factory where proteins are made on an assembly line. . . . To a large extent, it has been a mystery how this assembly line is laid out and these papers [reports of the experiment] show where different pieces of equipment are on the shop floor. . . . The [tRNAs] come through on a conveyor belt, and we can see how the ribosome holds the [tRNA] differently in each of the binding sites" (Stephens, 1999, p. 2).

The ribosome used in this research is found in the bacterium *Thermus thermophilus*. It has a mass of about 2.5×10^6 daltons, suggesting the presence of about 10^5 nonhydrogen atoms. The structure consists of 54 proteins and 3 RNA strands of a form known as ribosomal RNA (rRNA). The proteins and RNA are arranged into two subunits, the larger of which is called 50S and the smaller, 30S.

Researchers suggested that their work could be extended in two possible directions. First, improving the resolution of their map to the 3A level would allow the identification of individual atoms present in the protein and the construction of a precise atomic map of the ribosome. Second, further modifications of the research design might allow a continuous "motion picture" of the changes taking place over time on the ribosome. In the research reported here, maps were made of the ribosome when it was attached to some of the molecules involved at various stages of protein synthesis. These maps could be thought of as "snapshots" of certain steps in the manufacture of proteins, although most of the steps remain unknown. Researchers eventually are aiming for a way of filling in those missing gaps and obtaining, as the lead researcher described it, "a movie of the ribosome in action" (Stephens, 1999, p. 1). Should they be successful in reaching that goal, we would have the first clear images of a natural general assembler in operation, an important accomplishment in the field of molecular nanotechnology.

Another line of protein research involves the modification of natural proteins and the design and construction of synthetic protein-like molecules. The purpose of this research is to extend and diversify the kinds of proteins and protein-like molecules available and the kinds of functions they can offer. For example, natural systems make use of only four nitrogen bases in DNA molecules to encode information and 20 amino acids from which to construct proteins. Many more nitrogen bases and amino acids exist in nature, and even more can be made synthetically. The addition of even one more nitrogen base pair or one more amino acid to the natural "tool box" used by DNA and RNA would increase dramatically the number and diversity of proteins that could be formed.

Researchers at the University of California at Irvine reported in 1992 that they had made new mRNA and tRNA molecules capable of recognizing and manipulating amino acids not found in natural proteins. The work required the intervention of researchers at various points normally handled by natural molecules, but it at least showed that the 20-amino acid code used by natural systems is adaptable.

So much research has been conducted with regard to the elucidation and modification of protein structures that it is difficult to single out research of special significance. High on that list would be work on the structure of an entirely synthetic protein reported in 1999 by a research team headed by William F. De Grado of the Johnson Research Foundation at the University of Pennsylvania.

The goal of this research was to build an entirely synthetic protein, one not found in nature, that would have a specific, stable, and predictable three-dimensional structure. The work began with a review of existing proteins that might serve as a model for this synthetic analogue. De Grado's team selected a protein found in the bacterium *Staphylococcus aureus,* a common household germ. The team used a computer program to select a set of amino acids that would fold in the same way as its natural counterpart. Construction of the analogue proved successful in that the new protein did fold as predicted, forming a new protein that the team named $\alpha_3 D$. $\alpha_3 D$ is three times as large as any previously synthesized protein and has physical and chemical properties similar to those of natural proteins.

The next step in this research, De Grado points out, is to design some function into the protein. That is, an active site might be created to which metal ions, water molecules, or other proteins might bind. Those species then might become available for other steps in the construction of large molecules. De Grado says the ability to design and build synthetic proteins "takes us out of the realm of tinkering with existing proteins to engineering entirely new proteins and polymers" (Kreeger, 1999, p. 1).

Nucleic Acids

Nucleic acids are another promising material for the construction of nanosize machines and their components. As pointed out earlier, DNA molecules not only are stable, three-dimensional structures that function efficiently as nanomachines in living cells, but also they self-replicate naturally and spontaneously. Some researchers are exploring ways in which the properties of DNA can be used in the production of non-natural systems, such as the construction of nanosize devices.

One of the leaders in this field of research has been Nadrian Seeman at New York University. Seeman received the Foresight Institute's 1995 Feynman Prize in Nanotechnology for his work on DNA-based nanodevices. In his early research, Seeman explored the possibility of building rigid, three-dimensional objects from strands of DNA. His earliest successes were a DNA cube constructed from six different cyclic strands and a truncated octahedron consisting of six squares and eight hexagons. The goal of this research, according to Seeman, was "the rational synthesis of periodic matter and the assembly of a biochip computer" ("DNA Nanotechnology," n.d.)

In 1999, Seeman and colleagues reported on a new kind of DNA-based nanodevice, a robotic arm made of synthetic DNA. Figure 3.4 shows how this device works. Sections A and B of the device consist of *double-crossover* (DX) synthetic DNA molecules. The bonding within these molecules is strong enough to make both arms stiff. The two arms are joined by a short bridge of natural DNA in its "B" conformation. The B conformation of DNA is the configuration most commonly found in nature, the right-handed helical structure originally discovered by James Watson and Francis Crick in 1953. Under certain circumstances, however, DNA can occur in other forms. One of those is similar to the B form except that it forms a left-handed helix. That structure is known as the "Z" form of DNA. The conversion between the B and Z forms of DNA involves a winding or

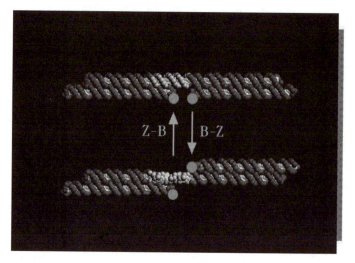

Figure 3.4. DNA robotic arm. *Courtesy of Dr. Nadrian Seeman.*

unwinding of the molecule by 3.5 turns. In the Seeman device shown here, the conversion between the two forms causes a shift in the relative position of the two arms.

To detect the relative movement of the two rigid arms, Seeman's team attached fluorescent dye molecules at their inner ends ("X" and "Y" in the diagram). The B↔Z transformation was detected by observing the movement of the two marker molecules relative to each other.

The marker-molecule feature of the experiment was important not only because of its role in the experiment, but also because of its potential use in future experiments. If marker molecules can be attached to the arms, so can other molecules, such as metal ions and proteins. The DNA robotic arm provides a method for moving these molecules from one position to another. Although this research provides a tantalizing hint as to the role of rigid DNA in nanomachines, Seeman cautions that it is "difficult to predict whether this system could also be used for a power nanoscale motor [or] estimate the ability of this system to position molecules precisely" (Mao, 1999, p. 145).

Basic Materials: Carbon Nanotubes

The growth, development, and confluence of nanoscale technologies toward the end of the 20th century was remarkable. Even if *Engines of Creation* had never been written, it is difficult to imagine that the potential for nanoscale devices would not have become obvious as a result of these technologies.

The need for new and improved pharmaceuticals with more efficient modes of delivery inspired a dramatic increase in the study of protein composition and folding. The challenge of designing and building faster computer elements at the submicron level forced researchers to explore the behavior of matter at the nanoscale level. The discovery of an unexpected new allotrope of carbon in 1985 opened up a new field of research with staggering potential for the production of nanoscale devices of low density and high mechanical strength and toughness. Developments in these three fields are expected to result in the manufacture of the smallest machines and electronic components ever made, perhaps of the first assemblers and replicators and almost certainly of the first nanocomputers.

The new allotrope of carbon was discovered in soot by a research team led by Harold W. Kroto at the University of Sussex in England and Richard E. Smalley at Rice University in Texas. These two men, and their co-worker at Sussex, Robert Curl, were awarded the 1996 Nobel Prize in Chemistry for this achievement.

The new allotrope of carbon consists of soccerball–shaped molecules made of 60 carbon atoms. The researchers named the C_{60} molecule *buckminsterfullerene* in honor of the famous American architect, visionary, and futurist Buckminster Fuller. Fuller is perhaps best known for the geodesic domes he used widely in his designs during the 1950s. A C_{60} molecule, similar to a geodesic dome, consists of a tightly packed combination of pentagons and hexagons, as illustrated in Figure 3.5.

Scientists quickly became intrigued by the modifications that could be made in a buckminsterfullerene molecule or, as it soon became better known, a *buckyball*. For example, the molecule could be cleaved along its diameter and a new ring of carbon atoms inserted between the two 30-carbon hemispheres. The process could be repeated over and over, producing an elongated tube of carbon atoms sealed at the ends by hemispheres of carbon atoms. These and other modifications of the basic buckyball became known as *fullerenes*.

In 1991, the Japanese electron microscopist Sumio Iijima made a remarkable discovery that revolutionized research on fullerenes. In his research, Iijima vaporized a sample of carbon between two electric arcs, then used an STM to analyze the soot formed. He found that the soot consisted of many long cylindrical tubes only a few nanometers in diameter but hundreds or thousands of nanometers in length.

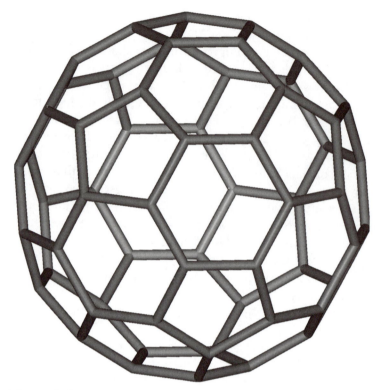

Figure 3.5.   Diagram of C_{60} molecule, known as a *buckyball*.
Courtesy of Dr. Richard Smalley, Rice University.

Iijima called these structures *carbon nanotubes*. The nanotubes were similar to those that researchers had been making by the painstaking method of slicing and splicing of buckyballs and their fragments. Iijima's research showed that those nanoparticles could be made simply and abundantly.

Researchers soon began to study the properties of nanotubes with an eye to their potential use in the construction of nanodevices. One reason for their enthusiasm was the realization that fullerene-based materials are likely to have some important chemical and physical properties similar to those of another carbon allotrope, diamond, the hardest natural substance.

Carbon nanotubes can be thought of as sheets of graphene rolled up into cylinders. Graphene is a naturally occurring form of carbon consisting of flat sheets of carbon atoms bonded to each other. Because all carbon electrons are used in the formation of these bonds, none are available to bond to adjacent layers. Two graphene sheets placed on top of each other slide smoothly over each other with little friction. This property is put to use in the lead (which is really graphite) in lead pencils. When the pencil point (the graphite) is rubbed against a piece of paper, some carbon easily rubs off the point and is deposited on the paper.

Graphene sheets have two significant physical properties. First, they have the highest tensile strength of any material known. If one can imagine a rope made out of carbon nanotubes, such a rope would be the strongest material ever produced, from 50 to 100 times stronger than steel. Second, the density of carbon atoms in a graphene sheet is greater than the density of any other two-dimensional material made of an element. Under normal circumstances, a nanotube or nanorope would be essentially impermeable.

An important step forward in research on carbon nanotubes occurred in 1993, when scientists learned how to make single-walled nanotubes (SWNTs). The first carbon nanotubes discovered in nature, such as those produced in Iijima's experiments, were multi-walled nanotubes (MWNTs). MWNTs consist of many concentric carbon cylinders nested inside each other. They are complex systems, more difficult to study than the simpler SWNTs. Using SWNTs, scientists rapidly learned a great deal about the electrical conductivity, tensile strength, flexibility, toughness, and other physical properties of carbon nanotubes.

Researchers at the University of Washington at St. Louis, the University of Wisconsin, and the Zyvex Corporation reported on the first measurements of the tensile strength of an SWNT. They first positioned an MWNT between two AFM cantilever probe tips and measured the force needed to cause the tube to fail, that is, to break. Analyzing the results of this experiment was difficult because the 19 different MWNTs used were of different diameters, ranging from 13 to 33 nm. Only the outer layer of each MWNT broke apart, leaving an inner shell that was greater in length than the original MWNTs. The authors of this paper decided to report only on the tensile strength of the broken outer layers, which they reported to be about from 11 to 63 gigapascals.

In another example of research on the properties of carbon nanotubes, two teams reported on the effects of adsorbed gases on the properties of SWNTs. The gases used were ammonia (NH_3) and nitrogen dioxide (NO_2) by one team and oxygen (O_2), by the other team. Researchers found that the two electronegative gases, NO_2 and

O_2, acted as p-type dopants, increasing the conductivity of the nanotubes. In some cases, the addition of O_2 converted the SWNTs from a semiconductor to a metal-like conductor. Ammonia seemed to have the opposite effect, essentially reducing the conductivity of the nanotubes to nearly zero.

Some of the most exciting discoveries in carbon nanotube research have come from the laboratory of Cees Dekker and colleagues at the Delft Institute of Technology in The Netherlands. In 1997, one of Dekker's teams showed that bent carbon nanotubes can function as electrical wires. Their electrical behavior is different from that of macroscopic wires, however. In the latter, each small increase in voltage produces a correspondingly small increase in current. The relationship between voltage and current is linear. In carbon nanotubes, however, current increases as a stepwise function of voltage. Increasing the voltage may or may not cause a current to flow.

The explanation for this phenomenon seems to depend on the number of electrons involved, an indication of the kinds of differences that may be observed in the behavior of matter at the macroscopic and nanoscale dimensions. In a macroscopic wire, billions of electrons are present. Any voltage increase results in the movement of at least some of those electrons. A carbon nanotube holds a much smaller and more limited number of electrons. The introduction of one or two new electrons into the nanowire is blocked by the electrons already there. Some given "push," that is, some increase in voltage, is required to force a new electron into the nanowire and an electron already present in the wire out, that is, in the flow of a small electrical current.

The importance of this discovery is that the stepwise relationship between voltage and current is precisely the kind of electrical behavior exhibited in some types of semiconducting devices used in computers. Carbon nanotubes hold the potential to serve as replacements for some semiconductors now used in mesoscale computer elements. Some of Dekker's other discoveries involving carbon nanotubes are discussed later in this chapter in sections dealing with molecular electronics.

Another major center of research on carbon nanotubes has been the Center for Nanoscale Science and Technology at Rice University. Director of the Center is Richard Smalley, a co-founder of the C_{60} allotrope of carbon. In 1996, Smalley and colleagues reported on the discovery of carbon "ropes" consisting of bundles of carbon nanotubes that self-organize into larger groups. The ropes were observed when a mixture of graphite and a nickel-cobalt catalyst were vaporized together by means of a laser beam.

When the product of this reaction was examined with an STM, single-walled nanotubes and carbon ropes were observed. The ropes consisted of 100 to 500 SWNTs, 10 to 20 nm in diameter and many microns long. Smalley pointed out that further research is needed to eliminate defects and reaction byproducts; to maximize yield; and to control tube diameters, length, chirality, and number of concentric shells. Nonetheless, according to one observer, the day may not be far off when molecular ropes can be produced in "continuous lengths that can be wound in a spool," a product that might have useful applications in the construction and operation of nanoscale devices ("Rice Reports Major Advance in Creating Carbon Nanotubes," *Foresight Update* No. 26, 1).

Some exciting results from research on carbon nanotubes using SPMs have been reported. The general paradigm for this research involves the use of one or more AFMs to move, bend, cut, twist, position, and otherwise manipulate carbon nanotubes, followed by an analysis of the ways in which these modifications have affected the electrical and other physical properties of the nanotubes.

A team of researchers at the University of Pennsylvania led by Alan Johnson reported in 1999 on their efforts to arrange carbon nanotubes into simple circuits on a silicon wafer. The researchers used an AFM to manipulate SWNTs on a surface of silicon dioxide. They developed the ability to slide the SWNTs around on the surface, lay them on top of each other, cut them apart, and sweep away unwanted fragments. The team was especially interested in finding out how the conductivity of the SWNTs was affected when one nanotube was placed on top of a second nanotube perpendicularly. Researchers attached metal contacts at the end of each tube, applied a voltage across it, and measured the flow of current. They discovered that the conductivity of the upper tube changed in the area where it was in contact with the lower nanotube. The kink formed in the upper tube at this point seemed to inhibit the flow of electrons, changing the tube from a conductor into a semiconductor.

These results suggest that it may be possible to arrange carbon nanotubes in such a way that they can perform the same functions as those currently found on a silicon chip. One commentator observed, "If methods can be found for generating such kinks in a controlled way . . . then we might look forward to microelectronics shrunk to molecular dimensions and based not on silicon but on carbon" (Ball, 1999, p. 120).

Other researchers are studying ways of making new nanostructures using carbon nanotubes. Jimmy Xu and colleagues at the University of Toronto have reported on efforts to construct Y-shaped structures made of carbon nanotubes. They first etched Y-shaped templates into a surface with a laser beam. Then they heated a mixture of acetylene (C_2H_2) and a cobalt catalyst to a high temperature over the surface. As the acetylene decomposed, carbon was formed and deposited as nanotubes in the grooves of the template. Researchers suggested that the Y-shaped carbon nanotubes produced might provide a mechanism for linking three other carbon nanotubes to each other, providing a method for joining such tubes not currently available.

The ultimate goal of much carbon nanotube research is to make nanoscale devices that can be used in nanocomputers, molecular machines, and perhaps assemblers and replicators. An example of the potential applications of such research was the manufacture of a telescoping bearing by researchers at the University of California at Berkeley (UCB) in 2000. The bearing consisted of two concentric layers of nanotubes with an outer diameter of about 8 nm. It was made by first attaching a nine-walled MWNT to a gold wire. A second rigid carbon nanotube attached to the tip of an STM was used as a tool with which to grasp the nine-walled MWNT.

Researchers used their STM tool to peel off the ends of the outer five layers of the MWNT. The resulting structure consisted of an inner tube of four layers (and an outer diameter of about 4 nm) protruding from the five-layer outer shell. The STM tool was welded to the inner tube, providing a way of pulling it in and out of the outer shell. Researchers observed the effects of this action with a transmission electron microscope capable of taking several images a second, providing a simple motion picture of the bearing's behavior.

The researchers observed that the inner tube slid in and out of the outer shell quite easily more than 20 times with no apparent friction or wear. The inner tube could be made to rotate within the outer tube, also with no apparent damage to either component. The device operated as a friction-free spring (the in-and-out mode) and as a friction-free bearing (the rotating mode).

Perhaps the most surprising result of the experiment was observed when the bond between the STM nanotube tool and the MWNT inner tube failed. When released from the tool, the inner tube contracted into its five-layer outer shell, apparently as the result of van der Waals forces between the inner and outer tube.

The UCB researchers completed their report by outlining its potential applications. "Our results demonstrate that multiwall carbon nanotubes hold great promise for nanomechanical or nanoelectromechanical systems (NEMS) applications. Low-friction low-wear nanobearings and nanosprings are essential ingredients in general NEMS technologies" (Sanders, 2000).

Applications: Molecular Electronics

Arguably the most powerful force driving nanoscale research today is the need for smaller, faster, and more powerful computing devices. Since the 1970s, scientists have been finding ways to cram more and more transistors—the basic unit of a computing device—onto a single chip. The simplest 4004 and 8080 processors in the early 1970s contained fewer than 10 transistors per chip. By 2000, that number had risen to nearly 10,000 transistors per chip in the Pentium II and III processors and to 100,000 transistors per chip in the Pentium III Xenon processor (Kaku, 2000).

This trend conforms to a pattern foreseen in 1965 by Gordon Moore, co-founder of the Intel Corporation. Moore predicted that the number of circuits on a silicon chip would double every year, a projection he later changed to a doubling every 18 to 24 months. Progress in chip design has followed Moore's Law with remarkable accuracy ever since. If that law holds true for the next two decades, one can expect processors with 1 billion transistors per chip sometime before 2015.

To meet this demand, engineers will have to find ways of reducing electronic components, such as wires and logic gates, to dimensions even smaller than those used in current devices, dimensions of about 100 nm. Present-day technology based on the use of intense laser beams for the construction of lithographic circuity is probably capable of achieving these goals. The problem is the potential costs involved. By some estimates, a chip fabrication facility needed to make these improved processors could cost more than $200 billion in 2015 (Reed and Tour, 2000). Leaders of the computer industry have questioned investments of that magnitude and have begun to explore alternatives to traditional solid-state computing devices. One of the most powerful of those alternatives is molecular electronics.

The term *molecular electronics* refers to the design and construction of electronic devices consisting of single molecules or small groups of molecules. By the turn of the 21st century, significant progress in this field already had been made. A review of some of those developments follows.

Molecular Photovoltaic Device

As is often the case, some researchers have turned to biological systems for models of possible synthetic molecular devices. The process of photosynthesis is a naturally occurring series of chemical reactions in which light energy is converted into electrical energy, then into chemical energy stored in chemical bonds. An example of the ways in which such natural systems might be modified was reported in 1997 by researchers at Arizona State University (ASU).

The ASU researchers studied a variety of synthetic molecular systems potentially capable of converting light energy into electrical energy, as occurs during photosynthesis. One such system consisted of a molecule containing three components: a carotenoid polymer (C), a porphyrin unit (P), and a fullerene (C_{60}) unit. The term *carotenoid* refers to any compound with a structure similar to that of carotene, a precursor of vitamin A with the chemical formula $C_{40}H_{56}$. A porphyrin is a nitrogen-containing compound containing four five-membered rings, often with a central metallic atom. This complex molecular structure can be represented symbolically as $C–P–C_{60}$.

When light is shone on this three-part system, the porphyrin unit donates an electron to the fullerene unit, forming a charged structure that can be represented as $C–P^{+}–C_{60}{}^{-}$. The carotenoid unit donates an electron to the porphyrin unit, forming a new structure, $C^{+}–P–C_{60}{}^{-}$, a system in which opposite ends of the molecule carry opposite electrical charges. A molecular system of this type might function as the first stage in a nanoscale device for generating electricity from solar energy, for example, as a nanophotovoltaic cell. Authors of the report on this research noted that their system "may help point the way to the development of molecular-scale (opto)electronic devices for communications, data-processing, and sensor applications" ("Molecular Electronics," 2001).

Molecular Circuity

One of the simplest electrical circuits that one could imagine would consist of a source of electrons; wires through which the electrons could flow; and a device for controlling the direction in which electrons flow, a diode or switch. The circuit could be made more useful by adding other elements, such as a device for storing information ("memory") or an appliance that could be operated by the flow of electrons. The first stages in the development of molecular electronics have involved research on the development of molecular wires and molecular diodes or switches.

Molecular wires

Two kinds of molecular wires have been proposed. One consists of long chains of organic molecules known as *polyphenylene* groups. Such wires are often known as *Tour wires* in honor of their inventor, James Tour, now at Rice University's Center for Nanoscale Science and Technology. Carbon nanotubes form the basis of a second type of molecular wire.

The first fully functioning molecular wire was described in early 1996 by a group of researchers headed by Tour, then of the University of South Carolina, and David Allara and Paul Weiss of Pennsylvania State University. These researchers used polyphenylene-like chains for their wires.

The basic unit of a polyphenylene chain is a benzene molecule. Benzene is a six-carbon ring with the molecular formula C_6H_6. When one hydrogen atom is removed, the resulting structure, $C_6H_5{}^\bullet$, is called a *phenyl* group. The unbonded position on the phenyl group (represented by the free electron,$^\bullet$, in the structure) is available for bonding with another atom or group.

When two hydrogen atoms are removed from the benzene molecule, the resulting structure, $^\bullet C_6H_4{}^\bullet$, is known as a *phenylene* group. The phenylene group may bond at both ends. Groups of phenylene groups may join to each other end-to-end to form polyphenylene, a long chain consisting of dozens or hundreds of phenylene monomers, as shown here:

$$-C_6H_4-C_6H_4-C_6H_4-C_6H_4-C_6H_4-C_6H_4-$$

Some of the electrons in this molecule are located above and below the plane of the molecule and are relatively free to move. An electrical potential applied at one end of the molecule causes electrons to flow through the molecule, producing an electrical current.

A polyphenylene chain can be modified in many ways to change its conductivity. For example, acetylene ($H-C\equiv C-H$) units can be introduced between phenylene groups to modify the flow of electrons, and saturated hydrocarbon units can be used as insulators around the phenylene chains to prevent leakage of electrical current from the chain.

A modification of the basic phenylene molecular chain was described by the Tour, Allara, and Weiss group in 1996. Researchers first attached a gold lead to a conducting surface. They then attached one end of a molecule derived from the polyphenylene molecule

[4,4'-di(phenylene-ethynylene)benzenethiolate] to the gold lead. To test the conductivity of this nanowire, researchers brought the tip of an STM into contact with the free end of the modified polyphenylene chain. When they did so, they observed a small flow of current through the gold-polyphenylene-STM circuit.

Carbon Nanotubes

An example of the use of carbon nanotubes as molecular wires was reported in mid-2000 by a research team headed by Hongjie Dai at Stanford University. The approach used in this experiment was to etch the surface of a material with a laser beam, lay down carbon nanotubes on the surface, and study the conductivity of the nanotubes as they were manipulated. Researchers heated a sample of methane gas in the presence of a metal catalyst over a surface of silicon dioxide (SiO_2). The methane decomposed, producing carbon and hydrogen, and some of the carbon condensed in the form of nanotubes on the SiO_2 surface. Dai's group attached metallic leads to opposite ends of individual nanotubes and passed currents through them to measure their conductivity.

An interesting result of this research was observed when researchers forced carbon nanotubes into valley-like depressions in the SiO_2 surface. They noted that the flow of current diminished sharply as the nanotube was bent with the tip of an AFM. This reduction in electron flow had not been predicted theoretically and contradicted an earlier experiment on the flow of electrical current in bent nanotubes. Dai's group explained this effect by assuming that carbon-carbon bonds formed between atoms on opposite inside walls of a nanotube at its greatest deformation, forcing a kink in the tube similar to that found in a kinked garden hose.

Researchers at the National Research Council of Canada (NRCC) reported in mid-2000 on another method of growing molecular wires. They began with a surface of pure silicon on which they deposited a thin layer of hydrogen, forming an unreactive covering of silicon hydride. Next, they used an STM to remove a single hydrogen atom from the surface, leaving behind a single silicon atom with a "dangling bond" (a pair of unbounded electrons). Finally, researchers exposed the surface to styrene gas. Styrene ($C_6H_5CH{=}CH_2$) is an organic molecule that polymerizes easily to form the familiar plastic known as *polystyrene*.

In the NRCC experiment, a single styrene molecule bonded to the pair of unbonded electrons on the silicon atom, forming a silicon-styrene complex. The formation of this bond caused a shift of electrons within the styrene molecule, exposing a new pair of unbonded electrons on one of the styrene's carbon atoms. These electrons then plucked a hydrogen atom from an adjacent silicon atom, creating a new dangling bond on that atom.

The process described here is the first step in a self-replicating process. Each step begins with the removal of a hydrogen atom from the silicon hydride surface (first by the STM and later by an activated styrene molecule) and ending with the removal of another hydrogen atom on an adjacent silicon atom.

Before long, a linear sequence of silicon-styrene complexes has been created. All that remains is the addition of an appropriate catalyst to encourage the reaction of adjacent styrene groups with each other, forming a long-chain polystyrene molecule. Although the polystyrene wires formed are not themselves conductive, researchers believe they could provide the basis for a hybrid device that is conductive, or they could serve as templates for the manufacture of conducting wires. The important feature of the research is that it provides a fast method of self-assembly by which large numbers of atomically precise chains can be constructed. As Bob Wolkow of the NRCC's Steacie Institute of Molecular Science said, "Right now it takes literally days to individually place atoms on a crystal using STM. With this method, you could use STM to prepare a batch of devices. Once exposed to the desired molecule, multiple lines would grow simultaneously. You could grow a virtually unlimited array of identical structures, all in a matter of seconds" (Wolkow, 2000, p. 2).

Molecular Diodes

A diode is an electronic device with two terminals that allows the flow of electrical current in only one direction. Researchers now are exploring the potential for two main types of diodes at the molecular level: molecular rectifying diodes (or molecular rectifiers) and molecular resonant tunneling diodes.

Molecular Rectifiers

The possibility of producing electrical rectifiers from single molecules or small molecular systems first was suggested in 1974 by Ari Avram at the IBM Thomas J. Watson Research Center and Mark A. Ratner at Northwestern University. Many researchers since then have attempted to produce molecules that would fulfill the conditions outlined by Avram and Ratner. The first two teams to design such structures successfully were led by Robert M. Metzger at the University of Alabama at Tuscaloosa and by Mark Reed at Yale University, both in 1997.

For their studies, Metzger's group used molecules of the compound hexadecylquinolinium tricyanoquinodimethanide. This molecule has three distinct parts, an electron donor region (D), an electron acceptor region (A), and a bridge (σ). In the hexadecylquinolinium tricyanoquinodimethanide molecule, the quinolinium segment—the double-ring structure at the left of the molecule—acts as the donor (D). The tricyanoquinodimethanide segment of the molecule, located at the right of the molecule, acts as the electron acceptor (A). The double bond (=) joining the two segments acts as a bridge (σ) through which electrons can flow.

This compound was tested as a rectifier in two ways. First, a layer one molecule thick was sandwiched between aluminum electrodes. Next, a multilayer sandwich of molecules was inserted between the electrodes. When a potential of about 1 volt was applied across each of these systems, a flow of current was observed from "left" to "right" but not from "right" to "left." That observation would suggest that the molecule had behaved as a rectifier, controlling the flow of electrons in one direction only between the two electrodes. Researchers concluded their report on this experiment by claiming that they had proved "that a monolayer of molecule 5 [the molecule described above] can rectify by intramolecular tunneling, and that monolayers and multilayers rectify both as macroscopic films and on a nanoscopic level" (Metzger, 1997, p. 10466).

The Reed group took a different approach to their design of a molecular diode. They first used a technique known as ballistic electron-beam microscopy to etch a hole only 30 nm wide in a nonconducting material. They allowed a group of about 1,000 Tour wires to self-assemble within that hole. Finally, the top and bottom of the self-assembled wires were coated with a thin conducting layer of metal. The top layer consisted of a gold-titanium alloy and the bottom layer consisted of gold metal. When voltage is applied to this system, current flows in one direction but not the other.

Molecular Resonant Tunneling Diodes

A resonant tunneling diode differs from a rectifier in one important regard: It permits the flow of current in either direction not, as with a diode, in one direction only. The operation of such a device is different from that of large-scale devices in that the laws of quantum mechanics are dominant. Consider a system similar to that in Figure 3.6. In this system, an electrical potential applied to the metal contact at the left of the diagram according to classic physics would not produce any flow of current through the system. The saturated hydrocarbon group ($-CH_2-$) would prevent that flow.

Under the laws of quantum mechanics, voltages of certain magnitudes applied to the contact can provide electrons with enough energy to tunnel through the energy barrier created by the hydrocarbon group. In such cases, current may flow from left to right through the circuit. If a comparable voltage is applied to the contact at the right of the diagram, current also can flow from right to left. The system shown here then acts as a switch that allows current to flow, provided that the applied voltage is of the correct magnitude.

The first molecular switch was announced by researchers at the University of California at Los Angeles (UCLA) and the Hewlett-Packard corporation in July 1999. In some regards, the device was not revolutionary because it was created essentially with existing "top-down" technology. Its crucial new feature was that it relied on individual molecules to perform a switching function.

Researchers began with a silicon chip coated with a layer of SiO_2. They used conventional technology to lay down a series of aluminum wires on top of the SiO_2. Next, they covered the whole surface with a single layer of rotaxane molecules. Rotaxane molecules are large, complex structures consisting of two or more independent segments not connected directly to each other, but linked through a linear piece threaded through a ring. Finally, a vapor containing titanium and aluminum was condensed on top of the rotaxane monolayer.

In essence, the system consisted of two metal electrodes (aluminum and aluminum/titanium) separated by a layer consisting of individual

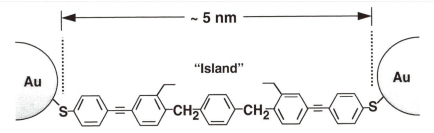

Figure 3.6. Molecular resonant tunneling diode. *J.C. Ellenbogen and J.C. Love. Architectures for Molecular Electronic Computers: 1. Logic Structures and an Adder Designed from Molecular Electronic Diodes. Proceedings of the IEEE, vol. 88, no. 3 (March 2000): pp. 386–426. Reprinted with permission.*

rotaxane molecules. In this configuration, the rotaxane molecules are able to conduct an electrical current. When a small voltage is applied, however, the rotaxane molecules undergo a chemical change, and they become nonconductive. The system qualifies as a switch because it is capable of allowing or preventing the passage of an electrical current. It is not a useful switch, however, because when the rotaxane molecules have undergone a chemical change, they cannot return to their original state. The system behaves as a one-time-only switch.

The UCLA scientists recognized that their rotaxane system was only the first step in the development of a more useful molecular switch. A year later, they reported on a redesigned switch that was not limited to a single use. The key to the modified switch was the use of catenanes, rather than rotaxanes, in the molecular device. Catenanes are organic compounds containing two interlocking ring structures that are not bonded to each other. They are joined mechanically, however, in a manner similar to the links in a chain.

For their redesigned switch, the UCLA scientists chose a catenane in which the two linked segments can take one of two positions relative to each other. In one of those positions, an electrical current is able to flow through the system (the switch is "on"). In the other position, current is unable to flow (the switch is "off"). In contrast to their original, rotaxane-based device, the catenane device can be switched between its "on" and "off" positions an indefinite number of times.

Shortly after the UCLA–Hewlett-Packard team announced its first rotaxane switch, scientists at Yale and Rice Universities described another approach to the construction of a molecular switch. In this design, a molecule consisted of three benzene units loosely bonded to each other. To the center benzene unit were attached a nitro and amine group. The presence of these groups made the structure highly susceptible to the presence of an electrical field.

In the state described here, an electrical current flows easily from one end of the molecule to the other. When a voltage is applied to one end of the molecule, the electron arrangement within the molecule is distorted, however, and it is no longer able to conduct a current. Without the voltage, the molecule is in an "on" position; with a voltage applied, the molecule is in an "off" position.

Molecular Memory Devices

Another essential part of any computing device is a memory unit, a material that can take two forms, such as "on" and "off" and "0" and

"1". In solid-state devices, this "on/off" alternative can be expressed by magnetizing or demagnetizing a tiny piece of iron. Only more recently have researchers begun to explore ways of using individual molecules or small groups of molecules to store data.

One of the first reports of a molecular memory device came in 1999 by Mark Reed and colleagues at Yale University. This research was an extension of their earlier work on molecular resonant tunneling diodes, described previously. They modified the three-benzene complex by removing the amine group from the middle benzene unit, leaving behind only the nitro group attached to the benzene ring. The nitro group has a strong affinity for electrons. When voltage is applied to one end of the modified benzene complex, its electron orbitals are strongly deformed, preventing the flow of electrons through the molecule. When the voltage is reduced, the orbitals become relaxed, and current flows through the molecule.

These two conditions can be used to represent "on" and "off" or "1" and "0" states of the molecule. That is, the molecule behaves as an information storage or memory device. One of the impressive findings of this research was the persistence of the molecule's condition. It was able to hold its state—either "on" or "off"— for nearly 10 minutes. That amount of time is remarkable because random-access memory units in a solid-state computer retain their state for only a few milliseconds and can be used for information storage only because they are constantly being refreshed by an external circuit.

Single Molecule Switch

One of the most basic components of any electrical or electronic device is a switch. A switch is a device that, in one position, allows electrical current to flow and, in a different position, prevents the flow of current. Switches are basic components of memory and logic systems in computers.

Any effort to reduce the size and increase the speed of computers depends, to some extent, on reducing the size of switches. At the present time, the smallest switch that one can imagine would consist of a single molecule. A molecular switch would be one-millionth the size of the smallest switch now used in transistors. In June 2001, a group of scientists led by James M. Tour of Rice University and Paul S. Weiss at Pennsylvania State University reported that they had prepared such a switch.

The switch consisted of (1) individual molecules of a phenylene ethynylene oligomer or, alternatively, (2) bundles of oligomer molecules. A phenylene ethynylene oligomer is a long molecule consisting of alternate phenylene (benzene rings) and ethynylene ($-C\equiv C-$) groups. The molecule contains single bonds around which groups within it can rotate, allowing at least two different spatial structures to occur. In one conformation, the ON position, the molecule allows an electric current to flow from one end to the other. In a second conformation, the OFF position, the molecule prevents the flow of the electric current. Thus far, researchers do not know exactly what conformations the molecule takes in its ON and OFF positions.

Molecules maintained one position or the other for varying lengths of time, ranging from a few seconds to tens of hours. One important factor determining the persistence of position was the surrounding matrix. When that matrix was well-ordered (as in a crystalline structure), molecules had a low tendency to change conformation. That is, they remained in an ON or OFF position for relatively long periods of time. When the surrounding matrix was more randomly arranged, molecules switched orientation more rapidly. An important goal of further research is to discover how the switching behavior in the molecule can be controlled.

Smallest Transistor

Modern computers were made possible in the late 1940s by the invention of the transistor. A transistor is an electronic device made of a semiconducting material, such as silicon or germanium, capable of acting as a switch or an amplifier. Attached to a transistor are three electrical contacts, the source, the collector, and the gate. In the absence of an electrical potential on the gate, the semiconducting material acts as an insulator, and no current flows through the device. Applying a potential to the gate alters the electrical properties of the transistor, however, allowing current to flow from source to collector.

Any effort to reduce the size of computers depends, to a large extent, on finding ways to reduce the size of transistors. In the fall of 2001, researchers at Lucent Bell Laboratories announced a major breakthrough in achieving this objective. They reported the preparation of a single-molecule transistor.

The basic problem in working at the molecular level is finding a way to manipulate molecules and small groups of molecules so that they can be arranged in some desired position. Lucent researchers solved this problem by allowing the target molecules to assemble themselves so as to form a transistor-like configuration.

They first etched a narrow notch into a thin gold layer deposited on a block of silicon. Next, they immersed the gold-covered block in a solution of an organic thiol, such as 4,4'-biphenyldithiol or 5,5'-terthiophenedithiol. These compunds are organic (carbon-containing) compunds with two or more thiol(–SH) groups. Thiols have a tendency to react with and form bonds with gold. When the block was removed from the organic thiol bath and allowed to dry, the gold surface had become covered with a single layer of organic thiol molecules.

The Lucent researchers then deposited a second thin layer of gold on top of the organic thiol molecules. The product was a "sandwich" with the two gold layers as the "slices of bread" and the organic thiol monolayer as the "filling." The two gold layers corresponded to the source and collector electrodes in a transistor and the organic thiol monolayer to the semiconducting material.

Finally, a third layer of gold was laid down at right angles to the first two gold sheets. This layer was to function as the gate electrode. When an electric potential was applied to the gold gate electrode, a flow of current was observed from the source electrode to the collector electrode through the organic thiol monolayer.

In its earliest form, this nanotransistor consisted of thousands of organic thiol molecules in the monolayer between the two gold electrodes. In a later refinement, however, researchers found a way of isolating single molecules of organic thiol from each other. They prepared this form of the nanotransistor by dipping the silicon block in a very dilute solution of organic thiol dissolved in an inert organic solvent consisting of an alkanedithiol, such as 1,5-pentanedithiol. Alkanedithiols are also organic thiols, but they do not conduct an electric current. When removed from this solution and allowed to dry, the gold surface was covered largely with inert, non-conductive molecules of the alkanedithiol, with small numbers of widely separated single molecules of the conductive organic thiol. When current was applied to the gate electrode in this nanotransistor, a flow of current was observed through the individual molecules of organic thiol, making them the equivalent of a single-molecule transistor. The research thus opens the way for designing and building newer and smaller computing devices made from what currently seems to be the smallest units possible, single-molecule transistors.

An Intramolecular Circuit

Computers carry out a myriad of mathematical functions using three basic kinds of electrical circuits, called *logic circuits*. These three circuits are the AND, OR, and NOT logic circuits. AND and OR logic circuits take digital data from two input electrodes and covert them to a single output. A NOT circuit, by contrast, simply reverses the input it receives. Thus a high voltage input to a NOT circuit (read as a "1") is converted to a low voltage output (read as a "0") and vice versa. Because of this action, a NOT circuit is also called an *inverter circuit*.

NOT circuits require two types of semiconducting materials, a p-type transistor and an n-type transistor. These two types of transistors differ from each other in the way electrical current is carried. In a p-type transistor, current is transmitted by the flow of positive "holes" in the material, while in an n-type transistor, current is transmitted by the flow of negatively-charged electrons.

One possible step in reducing the size of computers would be to find a way of making logic circuits from a small number of molecules or, ideally, from a single molecule. In August 2001, researchers at the IBM T. J. Watson Research Center in Yorktown Heights, New York, reported on the first single molecule NOT circuit. This circuit was made using carbon nanotubes especially modified for the purpose.

The first problem facing the IBM researchers is that all carbon nanotubes previously studied were p-type semiconductors. That is, it appears that electrical current is normally transmitted through carbon nanotubes through the flow of positive holes, not through the flow of electrons. In order to make a NOT circuit, then, it would be necessary to convert a naturally-occurring p-type nanotube into an n-type semiconductor. Methods for making this conversion had already been developed. For example, simply exposing nanotubes to the vapor of an electropositive element (potassium was commonly used) would achieve this objective. The IBM researchers found, however, that a simpler method for converting the character of a nanotube was available. By simply heating a nanotube in a vacuum, its character was changed from p-type to n-type. When the nanotube was exposed to air at atmospheric pressure, the process was reversed, and the n-type material reverted to its normal p-type configuration.

The first nanoscale NOT circuit made by the IBM team, then, consisted of two carbon nanotubes, one p-type and one n-type, attached through gold electrodes with conductive silicon as the gate electrode. This circuit showed voltage inversion, as required of a

NOT circuit, as well as current amplification of about three times. Current amplification is necessary in a logic circuit if data is to be passed successfully through a number of stages.

The next step in the IBM work was to make a NOT circuit out of a single carbon nanotube. To do so, researchers laid down a nanotube across two gold electrodes fixed to a silicon base. The array was then covered with a protective coating of a non-conductive material called PMMA. Finally, the coating above a short section of the nanotube was removed by means of electron beam lithography and the section was treated with potassium vapor. By means of this process, one portion of the nanotube (protected by the PMMA coating) had the character of a p-type semiconductor, while the exposed portion of the nanotube had the character of an n-type semiconductor. When voltage was applied to the gate electrode, the device acted as a NOT circuit, converting a high input voltage to low voltage, and vice versa. In addition, the single-molecule NOT circuit showed a voltage gain of 1.6, sufficiently high to make the device potentially useful in any more complex circuit.

Single Electron Transistor

The ultimate in miniaturization of electronic devices and systems requires the use of individual atoms, individual molecules, and even individual electrons. The concept of single-electron devices—devices through which electrons pass one at a time—is not new. In 1985, Dmitri Averin and Konstantin Likharev at the University of Moscow suggested the possibility of a single electron transistor (SET) made of two metallic electrodes separated by an insulator about 1 nm thick. Electrons would be able to flow through such a device by means of the quantum process known as "tunneling." During tunneling, electrons disappear from one electrode and re-appear at a second electrode without actually having traveled across the insulating barrier between them. Only two years after Averin and Likhaarev published their theory, the first SET was made by two researchers at Bell Laboratories, Theodore Fulton and Gerald Dolan.

With the rise of interest in nanoscale devices, researchers have been looking for ways in which small groups of molecules—even single molecules—can be made to function as SETs. In June 2001, researchers at the Delft University of Technology in The Netherlands working under the direction of Cees Dekker reported that they had produced such a device, a single-molecule SET. The SET consisted of a single carbon nanotube about 1 nm wide and 20 nm long. Researchers used an atomic force microscope to bend the nanotube in two places, producing a step-shaped drinking straw object. The kinks in the nanotube constricted the flow of electrons so that only one electron at a time could pass through the space between the two bends in the tube.

Researchers next attached two electrodes (the source and the drain), one at either end of the nanotube, and a third electrode (the gate) to the portion of the nanotube between the two kinks. When an electrical potential was applied to the gate electrode, a flow of electrons, one at a time, was observed across the nanotube from source to drain. Furthermore, the flow occurred in a very orderly fashion: when one electron flowed out of the middle section of the nanotube, across the kink and into the drain electrode, a second electron flowed out of the source electrode, across the kink, and into the central space between the two kinks in the nanotube.

An important feature of the Delft device was that it operated at room temperature. Previous SETs functioned only at very low temperatures, close to absolute zero, where the thermal motion of surrounding atoms did not mask the flow of single electrons. In the case of the Delft SET, however, the carbon nanotube is so small compared to surrounding materials that thermal interference with the single-electron electrical signal is insignificant.

Applications: Nanoscale Machines

One of the most striking trends in molecular nanotechnology has been the appearance of a variety of nanoscale devices that look as if they might be precursors of assemblers and replicators. These devices include nanoscale abacuses, pens, wheels, trains, motors, tweezers, and other machine-like gadgets. It is premature to predict that such devices confirm the possibility that assemblers and replicators can be built. Some of the scientists who have constructed the devices have rejected explicitly the possibility that Drexlerian nanomachines can *ever* be produced. Nonetheless, the appearance of these new nanodevices suggests at the very least that earlier cautions that molecular-sized devices could not be built were overly conservative. Following are descriptions of some of the nanoscale devices that have been produced.

Nanosize Abacus

One of the earliest nanoscale devices to be manufactured was a tiny abacus made with buckyballs. Researchers at IBM-ZRL announced their work on this device in November 1996. An abacus is a primitive method for calculating, thought to have been used first in Mesopotamia in about 3000 B.C. The instrument still is used widely in the Far East. The modern abacus is made of small wooden or plastic balls that can be slid up and down on metal rods. In its original design, the abacus consisted of balls that could be moved up or down along grooves on a flat surface.

The IBM researchers used the earlier design of the abacus as the model for their *nanoabacus.* The researchers first used laser beams to etch nanoscale lines on a copper surface. Next, they arranged groups of buckyballs (C_{60} molecules) along the grooves on the copper surface in a 10×10 array, as shown in Figure 3.7. Finally, they used the tip of an STM to slide individual buckyballs back and forth along the grooves to make calculations. A movie showing the movement of buckyballs is available online at http://www.zurich.ibm.com/pub/hug/PR/Abacus.

World's Smallest Pen and Plotter

Chad Mirkin at Northwestern University described his invention of the world's smallest pen as a case of "making lemonade out of

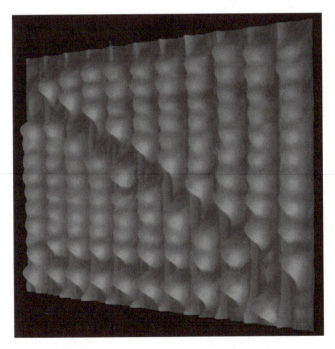

Figure 3.7. Molecular abacus. *Courtesy of IBM Zurich Research Laboratory.*

lemons." The lemons in this case were a result of using one of the most powerful scanning probe microscopy techniques, the AFM. When an AFM is used to probe the surface of a material, a tiny droplet of water tends to form on the tip of the instrument. The size of the droplet depends on the size of the AFM tip and the ambient moisture.

This problem is an annoyance researchers have to deal with, but to Mirkin and colleagues at Northwestern, it presented a new opportunity for "writing" at the nanoscale. Suppose, the researchers said, that the water-covered AFM tip is dipped into a colored compound. Then the voltage on the tip can be adjusted to release the colored water from the tip of the AFM onto a surface. Mirkin has called this new form of nanowriting *dip-pen nanolithography.*

The Northwestern team selected for its ink a compound known as octadecanethiol (ODT). ODT is an organic molecule consisting of 18 carbon atoms arranged in a long chain with two sulfur groups, one located at each end of the chain. The compound was chosen partly because sulfur tends to bond easily with gold, the surface (or "writing paper") on which the ink was to be laid down. When the AFM tip is dragged across the gold surface, molecules of ODT are deposited in a line only a few dozen molecules wide and one molecule thick.

Less than a year later, the Northwestern team reported on a further development of their nanopen, a nanoplotter. In the earliest version of the nanoplotter, researchers first laid down a thin line by the process described previously. They used the AFM in its imaging mode to see where the line had been deposited. This information was relayed to a computer that instructed the AFM as to where a parallel line some given distance from the first line was to be laid down. By this process, a series of parallel lines no more than 5 nm apart can be constructed.

Researchers experienced one problem with this approach. Because the AFM was being used in imaging and manipulation modes, Mirkin's team was concerned that during the imaging stage the AFM tip might displace some of the molecules that had just been laid down. Mirkin pointed out that such errors "could be unacceptable for electronic purposes and many other applications as it compromises the chemical integrity of the nanostructures" (Fellman, 1998).

To solve this problem, the Northwestern team modified the writing process slightly. First, they laid down grid marks on the gold surface using an ink consisting of the compound 16-mercaptohexadecanoic acid (MHA). MHA is an organic molecule with 16 carbon atoms arranged in a chain with a sulfur-containing group at one end. The MHA tip then was replaced with one containing ODT.

Next, the tip located the MHA reference lines and, using precalculated coordinates based on those lines, was used to draw three new lines parallel to and exactly 70 nm to the left of the MHA reference lines. Finally, the entire surface was imaged with an AFM whose tip was ink-free. The advantage of this approach, Mirkin pointed out, was that "[b]ecause the patterned lines are imaged only at the end of the process, cross-contamination of ink or molecules is prevented" (Fellman, 1998).

The Northwestern team expects its nanoplotter to become a "real workhorse for the nanotechnologist." Among its possible applications would be a master plate containing organic nanostructures designed to react with a variety of disease agents and circuits drawn with nanopens and nanoplots that could be used to test conductivity, thermal stability, chemical reactivity, and other properties of a material.

At the same time that the Northwestern researchers were working on their nanopen and nanoplotter, researchers at Michigan State University (MSU) and the University of Toronto were developing a second type of nanopen. This nanopen was similar to a fountain pen, compared with the quill-like pen developed at Northwestern.

In the MSU-Toronto research, a carbon nanotube was attached to the tip of an STM. The nanotube was dipped into a colored liquid and poised above the surface on which the ink was to be deposited. A pair of laser beams was aimed at the carbon nanotube, releasing electrons from the walls of the nanotube. The release of these electrons forced the ejection of ink molecules from the nanotube onto the writing surface.

Naturally Occurring Nanomotor

Serendipity has contributed to the development of molecular nanotechnology just as it has to most other fields of scientific research. The term *serendipity* refers to the unplanned and unexpected discovery of some new phenomenon or information. The discovery of the world's smallest motor is an example of such an event.

In 1998, scientists at IBM-ZRL, CNRS in Toulouse, and the Riso National Laboratory in Roskilde, Denmark, reported on the discovery of a natural molecular motor. The motor was found in a monolayer of hexa-*tert*-butyl decacyclene (HB-DC) laid down on an atomically clean copper surface. The HB-DC molecule consists of a central core made of 10 carbon atoms and the hydrogen atoms bonded to them. Attached to this central core are six legs made of t-butyl $[-C(CH_3)_3]$ groups. The legs project from the central core in such a way as to give the molecule a propeller-like structure.

Researchers used an STM to obtain an image of the HB-DC monolayer. The image essentially matched what they had expected: HB-DC molecules were arranged in a crystal-like pattern in which every molecule was held in place by other molecules surrounding it. Because of their shape, the molecules looked like a collection of hexagonal tiles neatly laid out on the copper surface.

The one surprising discovery made by the team was that interruptions of the tile-like pattern were observed occasionally. These interruptions consisted of blurred toroidal shapes in places where an individual HB-DC molecule had been expected. The image in such cases was similar to what one might expect if an individual HB-DC tile had been removed from the overall pattern and replaced by a smoke ring–shaped tile instead.

Researchers provided the following explanation for their observation: In some cases, an individual HB-DC molecule had been displaced slightly by a distance of about 1 nm from the position it would have had in a perfect tile pattern. When thus displaced, the molecule no longer was constrained by other HB-DC molecules around it. In such a case, thermal energy in the monolayer was sufficient to cause the displaced HB-DC molecule to begin spinning. The rate of spin was so rapid that the image obtained with the STM was that of a spinning propeller. Researchers argued that this explanation of their observation was supported by STM images and by theoretical calculations for changes of this kind. The practical significance of their discovery was that their results "open the way to fabricate, spatially define, and test recent proposals involving mechanical devices fabricated in molecular structures. They raise interesting questions concerning the fundamentals of mechanics in molecular and supramolecular systems, including the role of thermal noise and the design of molecular devices" (Gimzewski, 1998, p. 533).

World's Smallest Balance

One of the simplest and most elegant nanodevices yet created is a small balance. The balance was built by a research team headed by Walt A. de Heer at the Georgia Institute of Technology. The balance consisted of a carbon nanotube attached to the end of an SPM to form a cantilever. The research team was interested primarily in studying the properties of the carbon nanotube as it was bent. One main finding of the study was that the elastic mode of the nanotube changes as its diameter increases.

An interesting possible application of the device was in its use as a balance. An object placed at the end of the nanotube cantilever causes the tube to oscillate with an observable frequency. By knowing the properties of the nanotube, it is possible to determine the mass of the object from its pattern of oscillation. By using the cantilever probe in this way, the researchers were able to determine the mass of a tiny speck of soot to be about 22 femtograms. A femtogram is one quadrillionth (10^{-15}) of a gram, or one thousandth (10^{-3}) of a nanogram.

Nanotweezers

One of the basic tools needed for the construction of molecular devices would be an appliance that would allow an operator to pick up and manipulate parts for the device. In 1999, Philip Kim and Charles M. Lieber at Harvard University described their work on a pair of "molecular chopsticks"—they called them "nanotube nanotweezers."

The first step in constructing the nanotweezers was to deposit two fine gold electrodes on opposite sides of a thin glass micropipette. One set of bundled multiwall carbon nanotubes was attached to each of the electrodes. Nanotubes were selected not only because they are strong and stiff, but also because they conduct electricity. To operate the nanotweezers, the researchers applied a voltage to the two gold electrodes such that one electrode was positive and the other was negative. The nanotubes attached to the electrodes were charged similarly.

After the charge was applied, the carbon nanotubes on the two electrodes were attracted to each other with a force that depended on the voltage applied. At lower voltages, the two nanotubes bent toward each other but did not touch. At higher voltages, the two nanotubes came into contact with each other. By varying the voltage, the researchers were able to exert a fine degree of control as to how tightly the nanotweezers would close.

By using the nanotweezers in this way, Kim and Lieber could pick up various small objects, including a polystyrene sphere about 500 nm in diameter. Cellular components tend to be of approximately this size. The researchers also could pick up a single strand of gallium arsenide wire and remove it from a tangle of other wires.

The great advantage of nanotweezers, according to Kim and Lieber, are that they permit the manipulation of nanosize particles in three dimensions, not simply sliding them across a surface. Their next goal is to make nanotweezers even smaller in size so that they will be capable of picking up and manipulating individual atoms and molecules (Kim and Lieber, 1999, p. 2150).

Less than 1 year after the Harvard researchers had announced the construction of their nanotweezers, a second group of scientists at Lucent Technologies reported on another type of nanotweezer, this one about a thousand times smaller than the nanotube device described earlier. The Lucent nanotweezers are made of three strands of DNA, one of which (A) acts as a backbone on which the other two (B and C) are attached. The A strand is constructed to have a V shape with the B strand attached to one arm of the V and the C strand attached to the other arm. Both B and C strands extend beyond the ends of the A arms for a short distance. The total length of the nanotool is about 7 nm.

Under normal conditions, the Lucent nanotweezer remains in its open V shape. To operate the device, a fourth strand of DNA is brought into the vicinity of the nanotweezers. The fourth strand, described as being the *fuel* for the nanotweezers, is designed to be complementary to the B and C strands. As it approaches the nanotweezers, it forms bonds with both strands pulling them close to each other. Although the DNA nanotweezers are too small to be seen with any kind of microscope, their movement can be detected by placing fluorescent molecules at their ends. When the fuel strand is added to the nanotweezers, scientists can observe the two fluorescent spots slowly moving closer to each other.

The fuel DNA can be extracted from the nanotweezers by adding a fifth strand of DNA to the above-mentioned system. The fifth strand is designed to be complementary to the fourth strand and to bind more strongly with it than the fourth strand binds to the B and C strands. As the fifth strand removes the fourth strand, the nanotweezers again open up, and the fuel strand floats away as "exhaust fumes."

According to one of the Lucent researchers, the nanotweezers could have great value in nanotechnology research: "Of course, it's all very speculative, but you can imagine, for instance, little factories on chips doing chemistry or simple assembly. You can think of production lines made up of little motors with different reactants being passed from one place to the next" (Amos, 2000).

Synthetic Nanomotors

One of the nanodevices that scientists reasonably might be expected to try to build is a nanomotor, a device for converting solar, electrical, chemical, or some other form of energy into mechanical energy. Many kinds of nanomotors occur in living organisms, such as those found in muscle fibers, cilia, and flagella. Relatively little is known, however, as to how these bionanomotors operate, and the task of building their synthetic analogues is daunting.

Progress already has been made in this field, however. Two simple nanomotors were described in the September 9, 1999 issue of the journal *Nature*. One of these devices was constructed by a team of researchers under the direction of T. Ross Kelly at Boston College. For many years, the Boston College team has been constructing complex chemical structures with a variety of possible mechanical functions. They have made molecular brakes, molecular ratchets, and a primitive molecular motor. The design of these devices depends on the ability to invent and construct molecular components that operate much as their macroscopic counterparts do.

The molecular wheel announced in 1999 consists of two parts, one of which is a group of three benzene rings attached to each other in a three–paddle wheel structure called *triptycene* (Figure 3.8). The rings are designed to rotate around a central axis. The wheel is fixed within a surrounding stable structure consisting of a honeycomb-shaped plate made of a four-benzene structure called *tetracyclic helicene*. The complete molecular structure, wheel and housing, contains only 78 atoms.

The paddle wheel is caused to rotate within its casing by creating a new chemical bond between one of the paddles and the housing. The bond pulls the paddle wheel through an angle of 120 degrees before

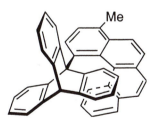

Figure 3.8. Diagram of a molecular wheel. *Courtesy of Professor Ross Kelly, Boston College.*

it comes to a stop. The energy needed to bring about this reaction comes from a compound known as carbonyl dichloride, a noxious gas also known as *phosgene.*

The Boston College nanomotor is the simplest possible device for converting chemical energy into mechanical energy because it travels a distance of only 120 degrees and no further. To get the nanomoter to move another 120 degrees, the whole process of creating another bond between paddle wheel and housing must be repeated. The goal of further research at the Boston College laboratory is to "optimize the system so that it rotates continuously and rivals the speed of its biological and mechanical counterparts" ("Professor T. Ross Kelly," 1999).

Although apparently only a modest step forward, Kelly's research has drawn kudos from some observers. For example, John Schwab, of the National Institute of General Medical Services, has called the work "significant from two points of view." In the first place, Schwab points out, the research will help us understand, "at least in some sense, how nature might convert chemical energy into controlled motion." In addition, the research goes "orders of magnitude beyond [top-down] nanotechnology, which we can visualize using optical microscopy or electron microscopy. We're taking it all the way to the single molecule level, which is quite exciting" (Machalek, 1999).

The second nanomotor described in *Nature* was developed by a team of Dutch and Japanese scientists. The nanomotor was discovered accidentally during the group's work on certain chiral molecules. Chiral molecules are molecules that have exactly the same chemical structure but different spatial orientations. They are precise mirror images of each other, just as are the two gloves that make up a pair.

Researchers had discovered that one molecule being studied could be converted from one form to another simply by shining ultraviolet light on it. When the molecule was heated to 60°C, it underwent a second transformation that brought it into a position 180 degrees different from its original position. In further experiments, they found that shining light on the molecule caused it to go through four distinct transformations, each involving a rotation of 90 degrees. In effect, by simply shining light on the molecule, they were able to make it rotate on its own axis continuously. The Dutch-Japanese nanomotor has the advantage over the Boston College motor of being smaller (58 atoms compared with 78 atoms) and of being able to operate continuously. One disadvantage is that the Dutch-Japanese nanomotor operates at a higher temperature than the Boston College device.

The lead author of the Dutch-Japanese paper, Ben Feringa, acknowledged that their invention is "a primitive system," but, "[n]ow we know it works, I'm convinced we can redesign it to rotate much faster" (Chang, 1999).

A third nanomotor built on different principles also was announced in 1999 by researchers at the Cornell University Department of Agricultural and Biological Engineering. These researchers combined biological and synthetic systems to create a nanomotor that ran for 40 minutes at a rate of three to four revolutions per second.

The fundamental unit of the hybrid device was a molecule of the adenosine triphosphatase (ATPase) enzyme (see Chapter 1). Large numbers of the molecule were produced synthetically by an *E. coli* bacteria culture that had been genetically modified. Molecules of the enzyme were attached to a substrate that had been prepared by traditional "top-down" lithographic techniques in which a precise pattern was etched on a glass plate with a high-energy electron beam. Tiny droplets of nickel, copper, or gold metal were laid down at specific positions on the pattern, and enzyme molecules were attached to the top of each metal droplet.

Finally, microspheres of a fluorescent material were attached to the free end of the enzyme arms at the top of the molecule, and a supply of ATP was provided to the system. As ATP was hydrolyzed by the enzyme, chemical energy was released, and the enzyme rotated counterclockwise, moving the fluorescent tip with it. The movement of the tip was easily observable by researchers, who were able to determine its rate of rotation. The important feature of this kind of nanomotor is its union of living and nonliving systems into a workable nanodevice. As lead author Carlo Montemagno observed, "The evolution of these technologies will open the door to the seamless integration of the motive power of life with engineered nanofabricated devices" (Montemagno and Bachand, 1999, p. 225).

Transportation Systems

One of the fundamental challenges in dealing with nanodevices is finding ways of moving them from one place to another. Scientists now are beginning to think about, plan, and design such systems. Two types of "train systems" have been proposed for the transport of nanoscale devices.

A theoretical proposal for one such system was described by Markus Porto and colleagues at Tel Aviv University in 2000. The system consisted of three nanosize groups of atoms joined to each other by spring-like bonds. The bonds are molecules of a photochromophore substance, a substance that shrinks or expands when exposed to light. The groups are laid down on a surface etched by traditional "top-down" lithographic methods to produce an egg-carton shaped pattern. The three particles making up any one system then nestle down into three adjacent depressions on the surface (Figure 3.9).

To move this tiny molecular "locomotive," a beam of light is shone on the system. The light causes the three-particle system to expand, causing one metallic cluster in the bundle to slide over into an adjacent depression. When the light is turned off, the bonds between metallic clusters shrink, pulling the second and third members of the

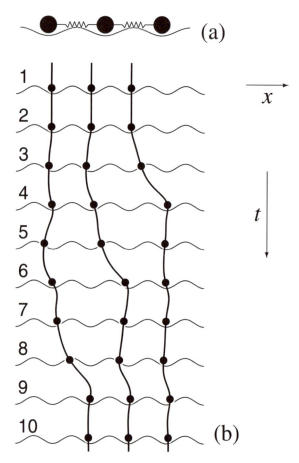

Figure 3.9.   Molecular train. *Courtesy of Professors Markus Porto, Michael Urbakh, and Joseph Klafter.*

cluster into adjoining depressions. By this mechanism, the complete cluster gradually creeps across the etched surface. Presumably, objects attached to the cluster (like the cars of a freight train) are carried along with the locomotive to new positions.

Another type of transport system has been suggested by Viola Vogel and colleagues at the University of Washington. Vogel's "railroad car" makes use of molecules found naturally in living systems to transport nanodevices. The key molecule in this system is *kinesin*. Kinesin is a protein capable of converting the energy stored in ATP molecules into mechanical motion. It occurs in nearly all cells and is responsible for transporting a variety of biological molecules from one place to another within the cell.

Here's how the process works: The kinesin molecule is a long, thin molecule that looks a bit like a stick-man. Projecting from one end of the molecule are two short legs and from the other end, two spherical heads. In cells, a kinesin molecule's legs attach themselves to organelles called *microtubules*, along which they walk when energy is supplied to them from ATP molecules. The walking behavior occurs a step at a time when energy from the ATP causes one leg to be released from the microtubule surface and move forward a distance of about 8 nm. When additional energy is available, the second leg is released from the microtubule, allowing it to move forward another 8 nm. By this process, the kinesin molecule slowly works its way along the organelle, carrying with it the load of molecules from one place to another.

Experiments have been done in the past using kinesin molecules to transport microtubules in nonliving systems. The first step in such experiments is to attach the heads of kinesin molecules to a glass plate. Microtubules then are added to the glass plate, where they are grabbed by the feet of the kinesin molecules and passed from molecule to molecule. This movement of microtubules is entirely random, however, and would not serve as a way of delivery the organelles (or anything else) in a predictable direction.

Vogel's adaptation of this experiment has been to etch the glass surface first by means of traditional "top-down" lithography or by using the newly developed dip-pen lithography developed at Northwestern University. In this way, precise, specific "tracks" can be laid down that restrict the position of kinesin molecules laid down on the glass surface. The kinesin heads all line up at the bottom of the grooves on the plate with their feet extended upward. Microtubules laid down on this surface will not be moved about randomly on the surface but instead will be passed from kinesin molecule to kinesin molecule in a specific direction determined by the shape of the etched lines. This system can be used for nanoscale transport if the devices to be carried (e.g., carbon nanotubes) can be designed to bond to kinesin molecules in a fashion similar to the way microtubules bond. Vogel sees great potential for this kind of transport system. "We are learning how to engineer a monorail on a nanoscale," she says. "We want a molecular shuttle that moves from point A to point B and which can be loaded and unloaded" (Knight, 1999, p. 41).

References

Amos, Jonathan. "DNA Makes Tiny Tweezers." *BBC News*, available at http://news6.thdo.bbc.co.uk/hi/english/sci/tech/newsid_873000/763097.stm.

Avram, A., and Ratner, M. A. "Molecular Rectifiers." *Chemical Physics Letters* 70 (1974): 277–83.

Ball, Philip. "Technology: It's a Small, Kinky World." *Nature* November 12 (1999): 119–20.

Binning, Gerd, and Rohrer, Heinrich. "The Scanning Tunneling Microscope." *Scientific American*, (August 1985): 50–56.

Brooks, Michael. "Drawing a Fine Line." *New Scientist* June 26 (1999): 11.

Brown, Chappell. "Chemical Researchers Design Molecular Computer." *EE Times*, available at http://www.eet.com/printable Article?doc_id=OEG1991109S0036.

Bumm, L. A., et al. "Are Single Molecular Wires Conducting?" *Science* March 22 (1996): 1705–07.

Cate, Jamie H., et al. "X-ray Crystal Structures of 70S Ribosome Functional Complexes." *Science* September 24 (1999): 2095–2104.

Chang, Kenneth. "The World's Smallest Motors." ABC News, available at http://abcnews.go.com/sections/science/DailyNews/nanomotors990908.html.

Chen, J. "Large On-Off Ratios and Negative Differential Resistance in a Molecular Electronic Device." *Science* November 19 (1999): 1550–52.

Collier, Charles P., et al. "A [2]Catenane-Based Solid State Electronically Reconfigurable Switch." *Science* August 18 (2000): 1172–75.

Collier, C. P., et al. "Electronically Configurable Molecular-Based Logic Gates." *Science* July 16 (1999): 391–94.

Culver, G. M., et al. "Identification of an RNA-Protein Bridge Spanning the Ribosomal Subunit Interface." *Science* September 24 (1999): 2133–35.

Cumings, John, and Zettl, A. "Low-Friction Nanoscale Linear Bearing Realized from Multiwall Carbon Nanotubes." *Science* July 28 (2000): 602–04.

Dennis, John R., Howard, Jonathan, and Vogel, Viola. "Molecular Shuttles: Directed Motion of Microtubules Along Nanoscale Kinesin Tracks." *Nanotechnology* September (1999): 232–36.

Derycke, V., et al. "Carbon Nanotube Inter- and Intramolecular Logic Gates." *Nano Letters* August 26 (2001). Available at http://pubs.acs.org/cgi-bin/jtextd?nalefd/asap/html/nl015606f.html.

"DNA Nanotechnology in Ned Seeman's Laboratory." Available at http://seemanlab4. chem.nyu.edu/nanotech.html.

Donhauser, Z. J., et al. "Conductance Switching in Single Molecules through Conformational Changes." *Science* June 22 (2001): 2302–2307.

Eigler, D. M., and Schweizer, E. K. "Positioning Atoms with a Scanning Tunneling Microscope." *Nature* April 5 (1990): 524–26.

Ellenbogen, James C., and Love, J. Christopher. "Architectures for Molecular Electronic Computers: I. Logic Structures and an Adder Built from Molecular Electronic Diodes." McLean, VA: The Mitre Corporation, 1999. This publication is a standard in the field of molecular electronics and is widely known as "the pink book." It is available online at http://www.mitre.org/ technology/nanotech.

Fellman, Megan. "Northwestern Chemists Plot the Next Step in Nanotechnology." *Northwestern News*, available at http://www.northwestern.edu/univ-relation . . . s/1998-99/*scimed/nanoplotter-scimed.html.

Gimzewski, James, et al. "Controlled Room-Temperature Positioning of Individual Molecules: Molecular Flexure and Motion." *Science* January 12 (1996): 181–84.

Gimzewski, J. K., et al. "Rotation of a Single Molecule within a Supramolecular Bearing." *Science* July 24 (1998): 531–33.

Hall, Alan. "Molecular Model-T." Available at http://www.sciam.com/ exhibit/1999/092099molecularmotor/index.html.

Henk, W., et al. "Carbon Nanotube Single-Electron Transistors at Room Temperature." *Science* July 6 (2001): 76–79.

Hong, Seunghun, and Mirkin, Chad A. "A Nanoplotter with Both Parallel and Serial Writing Capabilities." *Science* June 9 (2000): 1808–11.

Kaku, Michio. "What Will Replace Silicon?" *Time* June 19 (2000): 99.

Kelly, T. Ross, De Silva, Harshani, and Silva, Richard A. "Unidirectional Rotary Motion in a Molecular System." *Nature* September 9 (1999): 150–52.

Kim, Philip, and Lieber, Charles M. "Nanotube Nanotweezers." *Science* December 10 (1999): 2148–50.

Knight, Jonathan. "The Engine of Creation." *New Scientist* June 19 (1999): 38–41.

Koumura, R. W., et al. "Light-driven monodirectional molecular rotor." *Nature* September 9 (1999): 152–155.

Kong, J., et al. "Synthesis of Individual Single-Walled Carbon Nanotubes on Patterned Siliconwafers." *Nature* October 29 (1998): 878–81.

Kong, Jing, et al. "Nanotube Molecular Wires as Chemical Sensors." *Science* January 28; 622–25.

Koumura, Nagatoshi, et al. "Light-Driven Monodirectional Molecular Rotor." *Nature* September 9 (1999): 152–55.

Kreeger, K. Y. "Designer Molecules: Largest Protein Ever Created From Scratch Has Implications for Novel Drug Delivery and Diagnostics." Philadelphia: University of Pennsylvania Medical Center, 10 May 1999, p. 1

Lee, H. J., and Ho, W. "Single Bond Formation and Characterization with a Scanning Tunneling Microscope." *Science* November 26 (1999): 1719–21.

Lefebvre, J., et al. "Single-Wall Carbon Nanotube Circuits Assembled with an Atomic Force Microscope." *Applied Physics Letters* November 8 (1999): 3014–16.

Li, Jing, Papadopoulos, Chris, and Xu, Jimmy. "Nanoelectronics: Growing Y-junction Carbon Nanotubes." *Nature* November 18 (1999): 253–54.

Liddell, Paul A., et al. "Photoinduced Charge Separation and Charge Recombination to a Triplet State in a Carotene-Porphyrin-Fullerene Triad." *Journal of the American Chemical Society* February 12 (1997): 1400–05.

Lipkin, R. "System Breaks Individual Bonds." *Science News* June 24 (1995): 391.

Lopinski, G. P., Wayner, D. D. M., and Wolkow, R. A. "Self-Directed Growth of Molecular Nanostructures on Silicon." *Nature* July 6 (2000): 48–51.

Machalek, Alisa Zapp. "Mini-Motor Models Nature, Advances Miniaturization Technology." Available at http://www.nigms.nih.gov/news/releases/kelly.html. September 8 (1999).

Mao, Chengde, et al. "A Nanomechanical Device Based on the B-Z Transition of DNA." *Nature* January 14 (1999): 144–45.

Markoff, John. "Computer Scientists Are Poised for Revolution on a Tiny Scale." *New York Times* November 1 (1999), available at http://www.nytimes.com/library/tech/99/11/biztech/articles/01nano.html.

Metzger, Robert M., et al. "Unimolecular Electrical Rectification in Hexadecylquinolinium Tricyanoquinodimethanide." *Journal of the American Chemical Society* October 29 (1997): 10455–466.

"Molecule Makes Mini Memory." *TRN News*, Available at http://www.trnmag.com/Stories/081501/Molecule_makes_mini_memory_081501.html.

"Molecular Electronic Devices." Available at http://www.eng.yale.edu/reedlab/research/device/mol_devices.html. This is the home page for Mark Reed's laboratory group. It contains useful information on some of their research on molecular diodes, switches, and memory units.

"Molecular Electronics." Available at http://porphy.la.asu.edu/photosyn/faculty/gust/molecular_electronics.htm. 2001.

Montemagno, Carlo, and Bachand, George. "Constructing Nanomechanical Devices Powered by Biomolecular Motors." *Nanotechnology* September (1999): 225–31.

Phelps, Lewis M., "1995 Feynman Prize in Nanotechnology Awarded for Pioneering Synthesis of 3-D DNA Objects." *Foresight Update* No. 23, 1.

Piner, Richard D., "'Dip-Pen' Nanolithography." *Science*, January 29 (1999): 661–63.

Poncharal, Philippe, et al. "Electrostatic Deflections and Electromechanical Resonances of Carbon Nanotubes." *Science* March 5 (1999): 1513–16.

Porto, Markus, Urbakh, Michael, and Klafter, Joseph. "Atomic Scale Engines: Cars and Wheels." *Physical Review Letters* June 26 (2000): 6058–61.

"Professor T. Ross Kelly." Available at home page at http://chemserv.bc.edu/Department/faculty/kelly/kelly.html.

Reed, Mark A., and Tour, James M. "Computing with Molecules." *Scientific American* June (2000): 86–98.

Reed, M. A., et al. "Conductance of a Molecular Junction." *Science* October 10 (1997): 252–54.

"Rice Reports Major Advance in Creating Carbon Nanotubes." *Foresight Update 26*, September 15 (1996): 1.

Sanders, Robert. "UC Berkeley Physicists Create Tiny Bearings and Springs out of Carbon Nanotubes for Use in Microscopic Machines." Available at http://www.berkeley.edu/news/media/releases/2000/07/27_nano.html.

Schön, Jan Hendrik, Meng, Hong, and Bao, Zhenan. "Field-Effect Modulation of the Conductance of Single Molecules." *Science*, November 8 (2001): 2138–2140.

Schön, Jan Hendrik, Meng, Hong, and Bao, Zhenan. "Self-Assembled Monolayer Organic Field-Effect Transistors." *Nature*, October 18 (2001): 713–716.

Segelken, Roger. "Fantastic Voyage: Tiny Pharmacies Propelled through the Body Could Result from Cornell Breakthrough in Molecular Motors." Cornell News Service, available at http://www.cornell.edu/releases/Sept99/bio_nano_mechanical.hrs.html.

Shen, T. C., et al. "Atomic-Scale Desorption through Electronic and Vibrational Excitation Mechanisms." *Science* June 16 (1995): 1590–92.

Soreff, J. "Step toward nanotechnology." *Foresight Update*, No. 39 (1999): 10.

Stephens, T. "UCSC Researchers Obtain First Details [*sic*] Image of a Complete Ribosome." UCSC Public Information Office, September 20, 1999.

Tans, Sander J., et al. "Individual Single-Wall Nanotubes as Quantum Wires." *Nature* April 3 (1997): 474–77.

Thess, Andreas, et al. "Crystalline Ropes of Metallic Carbon Nanotubes." *Science* July 26 (1996): 483–87.

Tombler, Thomas W., et al. "Reversible Electromechanical Characteristics of Carbon Nanotubes Under Local-Probe Manipulation." *Nature* June 15 (2000): 769–72.

Walsh, Scott T. R., et al. "Solution Structure and Dynamics of a *de Novo* Designed Three-Helix Bundle Protein." *Proceedings of the National Academy of Science* May 11 (1999): 5486–91.

Weiss, Peter. "Atom Tinkerer's Paradise." *Science News* October 24 (1998): 268–69.

Wolkow, Bob. "NRC Scientists Grow Organic Wires for Nanoscale Devices." National Research Council of Canada press release, July 6, 2000, available at http://www.eurekalert.org/releases/nrcc-nsg070700.html.

Wolpert, Stuart. "UCLA Chemists Report Significant Progress Toward Molecular Computers; Research Ahead of Schedule." Available at http://www.uclanews.ucla.edu/Docs/1695.htm.

"The World's Smallest Abacus." Available at http://www.zurich.ibm.com/pub/hug/PR/abacus.

"World's Smallest Tweezers." *BBC News*, available at http://news6.thdo.bbc.co.uk/hi/english/sci/tech/newsid_557000/557388.stm.

Yao, Z., et al. "Carbon Nanotube Intramolecular Junctions." *Nature* November 18, 1999: 273–76.

Yu, Min-Feng, et al. "Strength and Breaking Mechanism of Multiwalled Carbon Nanotubes Under Tensile Load." *Science* January 28 (2000): 637–40.

Yurke, Bernard, et al. "A DNA-Fuelled Molecular Machine Made of DNA." *Nature* August 10 (2000): 605–608.

Chapter Four

Biographical Sketches

In the early 1990s, the number of scientists actively engaged in research on molecular nanotechnology probably could have been counted on the fingers of two hands. Today, that statement is no longer true. Researchers around the world have begun to explore the structure and properties of materials at the nanometer scale. Some of those researchers specifically acknowledge their familiarity with the goals of molecular nanotechnology and are working toward the development of assemblers and other nanoscale devices. Others may not agree that such goals are realistic. They may believe that the vision of "bottom-up" technology is unrealistic. Nonetheless, the work they are doing eventually may make some important contributions to the realization of that objective.

This chapter contains the biographies of some of the researchers who are working at the breaking edge of nanoscale research, whatever their personal views about molecular nanotechnology may be. The chapter also includes individuals who have been involved with the development of the Foresight Institute, the primary organization involved in educating the general public about a Drexlerian view of molecular nanotechnology. The list of researchers included in this chapter is not complete. So many people now are involved in the development of molecular nanotechnology, at one level or another, that some worthy individuals have had to be excluded from this chapter. The kinds of work being done in the field are well represented, however, by the biographical sketches that are provided.

Winners of the Foresight Institute's Feynman Prize in Nanotechnology are listed at the end of this chapter. The Feynman Prize first was given in 1993 for "recent work [that] has most advanced the development of molecular nanotechnology." The Prize was awarded in 1993 and 1995 to a single individual, but in 1997 and succeeding years it was divided into two parts, one for theoretical research and one for experimental research. The winners of this award are widely regarded as leaders in the field of molecular nanotechnology.

Phaedon Avouris (1945–)

Phaedon Avouris is currently manager of the Nanometer Scale Science and Technology group at the IBM T. J. Watson Research Center in Yorktown Heights, New York. He has worked extensively in the area of the manipulation and modification of materials at the nanometer scale. Avouris received the 1999 Feynman Prize for his research on carbon nanotubes for use in computing devices.

Avouris was born in Athens, Greece, in 1945. He received his B.S. from the Aristotelian University in Greece in 1969 and his Ph.D. in physical chemistry from Michigan State University in 1975. He spent 3 years in postdoctoral studies at the University of California in Los Angeles and the AT&T Bell Laboratories. In 1984, Avouris joined IBM, where he served as manager of the Chemical Physics group until 1994. He was then appointed to his current position.

Among Avouris' research interests have been ultrafast phenomena, surface physics and chemistry, proximal probe microscopy, and nanoelectronics. He currently is studying the structure and properties of carbon nanotubes and the relationship between electron structure and electron transport in nanostructures.

Avouris is a fellow of the American Physical Society, the American Vacuum Society, the American Association for the Advancement of Science, the New York Academy of Sciences, and the World Technology Network. In addition to the Feynman Prize, Avouris has received the Medard W. Welch Award of the American Vacuum Society and several IBM "Outstanding Technical Achievement Awards."

James C. Bennett (1948–)

James C. Bennett has been a director of the Foresight Institute since its founding in 1986. He is also a founding director of the Institute for Molecular Manufacturing.

Bennett was born in Tampa, Florida, in 1948. He was educated at the University of Michigan, where he majored in anthropology. Bennett has been interested in issues of high technology, international communications, and trade since 1978. In that year, he joined the Sabre Foundation's programs for international trade and space development. He was appointed head of Sabre's World Space Center, specializing in the organization of training and applications programs in space-related technologies for developing nations. As a result of this work, Bennett founded Free Zone Authority Services, Inc., a consulting agency specializing in trade zone management services.

In 1981, Bennett co-founded Space Enterprise Consultants, the first consulting firm devoted entirely to commercial space development. Subsequently, he was a co-founder of several space enterprises, including American Rocket Company, of which he became President in 1989. In 1992, he became a consultant to many businesses involved in the fields of space, communications, and other high technology projects. In 1997, Bennett founded Internet Transactions Transnational (ITTI), of which he is now President and Chairman.

Bennett has written and spoken widely about issues involving space commerce, high technology, nanotechnology, and related issues. He writes a weekly column, "The Anglosphere Beat," for United Press International and is the author of *The Anglosphere Challenge: The Future of the English-Speaking Nations Beyond the Internet Era* (Rowman & Littlefield, 2001).

Gerd Binnig (1947–)

Gerd Binnig shared the 1986 Nobel Prize for Physics with Heinrich Rohrer for their invention of the scanning tunneling microscope. The scanning tunneling microscope has become one of the most powerful tools for the analysis and fabrication of nanoscale devices and systems.

Binnig was born in Frankfurt am Main, West Germany, on July 20, 1947. He graduated from the Johann Wolfgang Goethe University in Frankfurt, then earned a Ph.D. in Physics from the University of Frankfurt in 1978. He accepted a position at the IBM Zurich Research Laboratory, where he and Rohrer developed their historic invention in 1981. In 1984, Binnig joined the IBM Physics Group in Munich. In addition to the Nobel Prize, Binnig has been awarded the German Physics Prize, the Otto Klung Prize, the Hewlett Packard Prize, and the King Faisal Prize.

Donald W. Brenner (1960–)

Donald W. Brenner currently is associate professor in the Department of Materials Science and Engineering at North Carolina State University (NCSU), where his special area of interest is computational materials science and new technologies for materials education. He has been actively involved with the Foresight Institute and served as co-chair of the organization's Eighth Conference held in Bethesda, Maryland, in 2000.

Brenner was born in Nyack, New York, on August 9, 1960. He received his B.A. in chemistry from the State University of New York at Fredonia in 1982 and his Ph.D. in chemistry from Pennsylvania State University in 1987. He worked as a research chemist in the Theoretical Chemistry Section of the U.S. Naval Research Laboratory from 1987 to 1994 before joining the Department of Materials Science and Engineering at NCSU. Among Brenner's research interests are atomistic simulation of chemical dynamics in condensed phases; structure, deposition, and properties of diamond films fullerene-based materials; and nanometer-scale fracture.

K. Eric Drexler (1955–)

K. Eric Drexler generally is regarded as the father of molecular nanotechnology. He first began thinking about this subject while he was a student at the Massachusetts Institute of Technology (MIT) in the early 1970s. At the time, many important breakthroughs in genetic engineering were taking place, and Drexler began to consider the possibility that all atoms and molecules, not just biological molecules, might be maneuvered in such a way as to build products "from the bottom up" rather than "from the top down." That idea since has formed the conceptual basis of the field of nanotechnology.

Drexler was born in Oakland, California, on April 25, 1955. He grew up in Lafayette, Indiana; New Haven, Connecticut; Cincinnati, Ohio; Denver, Colorado; and Monmouth, Oregon. In 1973, he entered MIT, from which he eventually earned his S.B. in Interdisciplinary Science in 1977.

When he entered MIT, Drexler was interested primarily in the subject of space travel and the possibility of establishing human civilizations in space. He eventually became a part of an informal group of students, faculty, and others from a variety of academic institutions interested in this topic. While still a freshman at MIT, he gave a presentation at the First Princeton Conference on Space Colonization, sponsored by Dr. Gerard K. O'Neil, one of the world's foremost authorities on the subject.

Drexler was (and is) not the kind of person to limit his interests to a single topic. As he thought about the possibilities and potentials for space research, he began to see ways in which microfabrication of materials could be useful in the development and production of essential materials and structures. Before long, this line of thought had led to the realization that an entirely new method for producing materials might be possible, a method that involved constructing objects beginning with individual atoms and molecules placed carefully in exactly the correct places.

During his days at MIT, Drexler continued to develop his ideas about nanotechnology and discuss and debate these ideas with other students and faculty. He did not publish anything on the subject, however, until 1979. In that year, he first found out about Richard Feynman's talk on "There's Plenty of Room at the Bottom" given 20 years earlier. In that talk, Feynman had outlined many of the essential ideas about which Drexler had been thinking and talking over the previous 6 years. He knew it was time to write a paper in which he laid out his thoughts about an entirely new field of technology—nanotechnology.

That paper, "Molecular Engineering: An Approach to the Development of General Capabilities for Molecular Manipulation," was published in the *Proceedings of the National Academy of Sciences* in September 1981. The paper drew virtually no attention and was not even cited apparently until 2 years later. Nonetheless, it had laid down the general principles of a new approach to technology and a new field of research that had the potential to revolutionize life on Earth.

In September 1979, Drexler received his S.M. degree in engineering from MIT. He accepted a position as research affiliate, first at the Space Systems Laboratory and, later, at the Artificial Intelligence Laboratory. In May 1985, he and his wife Chris Peterson left for California. The group with which he had been working on nanotechnology believed that California would be the most congenial place in which they could continue meeting and working on their ideas for this new form of technology.

Shortly after his arrival in California, Drexler was offered a position as visiting scholar in the Stanford University Department of Computer Science. He held that post until 1991, when he and his colleagues established the Institute for Molecular Manufacturing (IMM) in Palo Alto. Since 1991, Drexler has carried out research on nanotechnology at IMM, and has written two major books and many articles and technical papers on the subject for technical and general publications. In 1991, he was granted his Ph.D. from MIT in molecular nanotechnology, the first degree of its kind to be awarded anywhere.

Donald M. Eigler (1953–)

Donald M. Eigler is perhaps best known for his 1989 project in which he arranged 35 xenon atoms to spell out the IBM logo on a smooth copper surface. Eigler used a scanning tunneling microscope at temperatures near absolute zero to manipulate and place the atoms in their proper position.

Eigler was born in Los Angeles, California, on March 23, 1953. He received his bachelor's degree in physics in 1975 and his Ph.D. in physics in 1984, both from the University of California at San Diego. He did his postdoctoral studies at the AT&T Laboratories before joining IBM as a research staff member in 1986. In 1993, Eigler was promoted to IBM Fellow, the highest technical position in the corporation.

Eigler has been awarded the Grand Award for Science and Technology from *Popular Science* magazine (1990), the 1993–94 Newcomb Cleveland Prize from the American Association for the Advancement of Science, and the Dannie Heineman Prize (1995) from the Göttingen Academy of Sciences. In 1999, Eigler received the first Nanoscience Prize during the Fifth International Conference on Atomically Controlled Surfaces, Interfaces, and Nanostructures.

Robert A. Freitas, Jr. (1952–)

Robert A. Freitas, Jr., is author of *Nanomedicine*, a three-volume work on the applications of molecular nanotechnology to the field of medicine. The first volume of the projected series was published in October 1999.

Freitas was born on December 6, 1952, in Camden, Maine. He earned his B.S. at Harvey Mudd College in physics and psychology in 1974. He then attended the University of Santa Clara School of Law, from which he received his J.D. degree in 1978.

After graduation, Freitas became interested in astronomical and space research. From 1977 to 1982, he was editor and publisher of *Space Initiative*, a handbook advocating political actions related to the space program. From 1979 to 1984, he was a researcher in astronomy and SETI (search for extraterrestrial intelligence) programs at the Leuschner and Kitt Peak Observatories and the Hat Creek Radio

Observatory. From 1976 to 1986, he was a freelance science and technology writer for many popular publications, including *Omni*, *Science Digest*, *Technology Illustrated*, *Student Lawyer*, and *The Humanist*. Between 1987 and 1997, he was editor and publisher of *Value Forecaster*, a monthly economics and investment research newsletter.

Freitas has been interested in molecular nanotechnology since the early 1990s. In 1994, he was appointed research fellow at the Institute for Molecular Manufacturing, where he began work on *Nanomedicine*. In 2000, Freitas accepted the post of Research Scientist at Zyvex LLC, where he will complete the last two volumes of the *Nanomedicine* series.

Richard Feynman (1918–86)

Richard Feynman sometimes has been called the intellectual father of nanotechnology. This honor is based on a speech he gave at a December 1959 meeting of the American Physical Society in Pasadena, California. That speech was entitled "There's Plenty of Room at the Bottom" and outlined some of the general arguments for beginning research at the nanometer level. At the time, Feynman was unable to imagine any applications for such research and was sufficiently convinced that it would not take place in the near future to offer two $1,000 prizes for anyone who could produce certain nanolevel objects.

Feynman was born in New York City on May 11, 1918. He attended the Massachusetts Institute of Technology (MIT), from which he received his bachelors degree in 1939. He went on to earn a Ph.D. in physics at Princeton University in 1942. During World War II, Feynman worked on the Manhattan Project in Los Alamos, where he was in charge of a group working on problems of separating uranium isotopes by means of diffusion.

After the war, Feynman accepted an appointment in the Department of Physics at Cornell University. In 1950, he took a similar position at the California Institute of Technology, where he remained until his death in Los Angeles on February 15, 1988 after an 8-year battle with abdominal cancer.

Feynman's special area of expertise was the study of quantum electrodynamics. He was awarded a share of the 1965 Nobel Prize in Physics for his work in this area.

Feynman was widely known as an outstanding teacher and interpreter of science and technology for the general public. His *Feynman*

Lectures on Physics (in three volumes, with R. Leighton and R. Sands, 1963) is still one of the finest general introductions to the subject available. Feynman also wrote two enormously successful popular books about physics, *Surely You're Joking, Mr. Feynman* (W. W. Norton, 1985) and *What Do You Care What Other People Think?* (W. W. Norton, 1988). He probably is best remembered by many people for his work on the commission appointed to study the cause of the *Challenger* space disaster in 1986. He was the commission member selected to appear before television cameras to explain the O-ring failure that caused that disaster.

M. Reza Ghadiri (1959–)

M. Reza Ghadiri received the 1998 Feynman Prize in Nanotechnology for Experimental Work. The prize was awarded for Ghadiri's research on the construction of molecular structures by means of self-organizing units.

Ghadiri was born on December 8, 1959, in Teheran. He earned his B.A. from the University of Wisconsin at Milwaukee in 1982 and his Ph.D. in chemistry from the University of Wisconsin at Madison in 1987. He spent 2 years in a postdoctoral program from 1987 to 1989 at the Rockefeller University in New York City. He was assistant professor at Scripps Research Institute in La Jolla, California, from 1989 to 1994. Currently, Ghadiri is associate professor in the Departments of Chemistry and Molecular Biology at the Scripps and a researcher at the Skaggs Institute for Chemical Biology at Scripps.

Ghadiri has published more than 36 papers on the self-reproduction of polypeptides and has been granted three patents on metallopeptides. Ghadiri has been an invited lecturer at conferences, meetings, and other presentations in Japan, Mexico, Canada, Iceland, Germany, Spain, Switzerland, and Italy and throughout the United States. He was awarded the 1995 American Chemical Society Award in Pure Chemistry. His research interests include the *de novo* design of functional artificial proteins, peptide nanotubes, and related biomaterials; artificial transmembrane ion and molecular channels; design of biosensor arrays; construction of self-replicating molecular structures; and the study of early events of protein folding.

James Kazimierz Gimzewski (1951–)

James Kazimierz Gimzewski currently is a group leader at the Zurich Research Laboratory of the IBM company. Since the mid-1980s, he has been interested in research at the nanoscale level. He has pioneered research on electrical contacts with single atoms and molecules and was one of the first people to image molecules with a scanning tunneling microscope. Among his most recent discoveries are single molecule rotors, the fabrication of molecular suprastructures at room temperature, and the development of new types of micromechanical sensors.

Gimzewski was born in Glasgow, Scotland, on December 20, 1951. He attended the University of Strathclyde in Glasgow, where he earned his B.Sc. in pure chemistry, with honors, in 1974. He completed his Ph.D. in physical chemistry at Strathclyde in 1977. Gimzewski worked as a postdoctoral student at Oregon State University (1977–79) and the University of Zurich (1979–83) before accepting an appointment with IBM in September 1983.

Among Gimzewski's many honors and awards are the 1997 Feynman Prize in Nanotechnology (Experimental), the 1997 Discover Award for Emerging Fields (Nanotechnology), the 1997 IBM Fifth Invention Achievement Plateau Award, and recognition in 1997 as the "founder" of the United Kingdom Institute of Nanotechnology. Gimzewski is author of more than 150 papers, reports, and other publications and holder of 17 European and U.S. patents. He also has been speaker at more than 200 invited talks throughout the world.

Al Globus (1952–)

Al Globus has been actively engaged in research on molecular nanotechnology, especially with regard to its applications to space research, for nearly a decade. He was co-recipient of the 1997 Feynman Prize in Nanotechnology (Theoretical) and was co-chair of the Fifth and Sixth Foresight Conferences on Molecular Nanotechnology.

Globus was born in Chicago on May 13, 1952. He grew up in Southern California and Germany and reports that his "first career" was music, in particular, playing guitar bass and saxophone in rock, jazz, country-rock, and funk bands. After attending the Berkeley School of Music in Boston, he entered the University of California at Santa Cruz, from which he received a B.A. in information science in 1979.

Globus' first job after leaving Santa Cruz was with Informatics Inc./Sterling Software, Inc., at the NASA Ames Research Center (1979–88). He then joined CSC, Inc., at Ames before moving to a position with MRJ Technology Solutions, Inc., also at Ames. In April 2000, he returned to work for CSC, Inc., at Ames.

Globus is the author of about 36 papers and research reports and has been honored with the 1998 NASA Public Service Medal, the 1998 Ames Contractor Council NASA Team Contractor Certificate of Excellence, and the 2000 NASA Ames Group Achievement Award to the Integrated Product Team on Devices and Nanotechnology.

William A. Goddard, III (1937–)

William A. Goddard, III, was awarded a share of the 1999 Feynman Prize of the Foresight Institute for theoretical work in nanotechnology. He received the award because of his work on the modeling of molecular machine designs. Goddard shared the prize with colleagues Tahir Cagin and Yue Qi at the Materials and Process Simulation Center at the California Institute of Technology.

Goddard was born on March 29, 1937, in El Centro, California. He received his B.S. in engineering from the University of California at Los Angeles in 1960 and his Ph.D. in engineering science from the California Institute of Technology (Caltech) in 1964. He has spent his entire academic career at Caltech, as assistant, associate, and full professor of theoretical chemistry (1964–78), professor of chemistry and applied physics (1978–84), and director of various National Science Foundation projects (1985–97). From 1984 to the present, he has been Charles and Mary Ferkel Professor of Chemistry and Applied Physics. He also is director of the Materials and Process Simulation Center of the Beckman Institute at Caltech.

Goddard's research interests cover a variety of fields, including the use of force fields to describe the dynamics of atomic motions, the molecular dynamics of large molecules and solids, the use of parallel computer systems to study such behaviors, and the applications of these methods to problems in chemistry, biology, and material sciences. Among Goddard's many honors are the American Chemical Society Award for Computers in Chemistry; the Badger Prize for Instruction in Chemistry at Caltech; and election as a fellow to the National Academy of Sciences, the American Physical Society, and the American Association for the Advancement of Science.

John Storrs Hall (1954–)

John Storrs Hall has been involved in the study of nanocomputers for more than a decade. He is perhaps best known for developing the concept of *utility fog*, a kind of all-purpose molecular material that can be directed to take on virtually any shape that one wishes. He also has been moderator of the online *sci.nanotech* news and discussion group for 10 years.

Hall was born in Atlanta, Georgia, in 1954. He graduated with a B.A. *cum laude* from Drew University in 1976, then earned an M.S. and Ph.D. in computer science from Rutgers University in 1994. From 1973 to 1976, he was first a systems programmer at the Drew University Computer Center, then a research assistant in the university's Department of Computer Science. Between 1980 and 1998, Hall was a systems programmer, then a computer systems architect at the Laboratory for Computer Science Research at Rutgers. In 1998, he was appointed a research fellow at the Institute for Molecular Manufacturing in Palo Alto, California. Hall is author or co-author of more than a dozen articles and patents in computer science.

Jan H. Hoh (1959–)

Jan H. Hoh has long been active in annual Foresight Institute conferences. He presented the tutorial on molecular nanotechnology at the Sixth Conference and served as co-chair for the Seventh and Eighth Conferences.

Hoh was born in Stockholm, Sweden, on December 3, 1959. He attended Illinois State University, from which he received his bachelors degree in biology and chemistry in 1983. He then attended the California Institute of Technology, where he earned his Ph.D. in cellular biology and biophysics in 1991. He spent the period from 1991 to 1994 in two postdoctoral programs, one at the University of Basel and one at the University of California at Santa Barbara. In 1994, he was appointed assistant professor of physiology at Johns Hopkins University School of Medicine. He holds a joint appointment at Johns Hopkins in the Department of Chemical Engineering.

Hoh's special area of interest involves work with the atomic force microscope, a field in which he has published about 36 papers. He was invited to give a plenary lecture at the United Kingdom Scanned Probe Microscopy meeting and the Hascoe Distinguished Lecture in Nanosciences at the University of Connecticut, both in 1999.

Neil Jacobstein (1954–)

Neil Jacobstein currently is president and chief operating officer of Teknowledge Leadership, an artificial intelligence and Internet software company founded in 1981. He also has served as volunteer chairman of the Institute for Molecular Manufacturing (IMM) since 1992.

Jacobstein was born in Miami Beach, Florida, on October 20, 1954. He earned his B.S. *summa cum laude* in environmental science from the University of Wisconsin and M.S. in human ecology from the University of Texas. He worked as a graduate research intern in the Learning Research Group at Xerox PARC and as a consultant in PARC's Software Concepts Group. He also worked as an environmental research associate at Washington University in St. Louis and New York City before joining Teknowledge.

Jacobstein led the development of the Foresight Institute's Guidelines for Molecular Nanotechnology and moderated the 1999 Foresight/IMM Monterey Workshop that led to the first draft of the guidelines. He was appointed a Henry Crown Fellow at the Aspen Institute in 1999 and gave a talk on molecular nanotechnology to an audience of 1,500 at the Institute's 50th anniversary celebration.

James B. Lewis (1945–)

James B. Lewis is currently webmaster for the Foresight Institute website and works on website design and construction and as a consultant on technologies leading to the development of molecular manufacturing and nanotechnology. He was previously senior research investigator in the Immunodeficiency and Immunosuppression Department of the Bristol-Meyers Squibb Pharmaceutical Research Institute (1988–96), associate member in the Basic Sciences Division of the Fred Hutchinson Cancer Research Center (1980–88), and staff investigator and senior staff investigator at the Cold Spring Harbor Laboratory.

Lewis was born in York, Pennsylvania, on December 14, 1945. He received his B.A. from the University of Pennsylvania in 1967 and M.A. and Ph.D. in chemistry from Harvard University in 1968 and 1972. He did postdoctoral research at the Swiss Institute for Experimental Cancer Research in Lausanne from 1971 to 1973 and at Cold Spring Harbor from 1973 to 1974. Lewis has published about 50 scholarly papers and is co-editor of two standard texts on nanotechnology, *Nanotechnology: Research and Perspectives* (1992) and *Prospects in Nanotechnology: Toward Molecular Manufacturing* (1995).

Ralph C. Merkle (1952–)

Ralph C. Merkle is one of the most prominent researchers currently involved in molecular nanotechnology. He is at least as well known to the general scientific community for his role in the development of public key cryptography.

Merkle was born on February 2, 1952, in Berkeley, California. He attended the University of California at Berkeley, from which he received his B.A. and M.S. in computer science in 1974 and 1977. He then attended Stanford University, from which he earned his Ph.D. in electrical engineering. The topic of his doctoral thesis was "Secrecy, Authentication, and Public Key Systems."

After graduating from Stanford, Merkle accepted a position of manager of Compiler Development at the Elxsi Corporation. He held that post until 1988, when he was appointed research scientist at Xerox PARC in Stanford. In 1999, Merkle left Xerox to become principal fellow at Zyvex, the first molecular nanotechnology development firm.

While maintaining his interest in public key cryptography, Merkle also devotes much of his time to molecular nanotechnology. He is an advisor to the Foresight Institute, a senior research associate at the Institute for Molecular Manufacturing, and a member of the Executive Editorial Board for the journal *Nanotechnology*.

Merkle was awarded the Kanellakis Award for Theory and Practice of the ACM (Association for Computing Machinery) for his work on public key cryptography and was awarded the 1998 Feynman Prize for molecular nanotechnology theory. Merkle's home page (www. merkle.com) is one of the most valuable resources for additional information on his own work and on the field of molecular nanotechnology in general.

Marvin Minsky (1927–)

Marvin Minsky is a member of the Foresight Institute Board of Advisors and has long been one of the strongest supporters of molecular nanotechnology. He is quoted as saying that he was "always interested in nanotechnology, even as a child. . . . I was always intrigued by the little machines one studied in organic chemistry."

Minsky was born in New York City in 1927. He attended the Fieldston School and the Bronx High School of Science in New York City and the Phillips Academy in Andover, Massachusetts. After a short stint with the U.S. Navy (1944–45), Minsky entered Harvard University, from which he received his B.A. in 1950. He continued his studies at Princeton University, from which he earned a Ph.D. in mathematics in 1954.

After holding the position of junior fellow at Harvard from 1954–57, Minsky accepted a job at the M.I.T. Lincoln Laboratory. He then became assistant professor of Mathematics (1958), co-director of the Artificial Intelligence Laboratory (1959–74), professor of electrical engineering (1974), Donner Professor of Science (1974–89), and Toshiba Professor of Media Arts and Sciences (1990–present), all at M.I.T.

Minsky also founded the M.I.T. Artificial Intelligence Project in 1959, a field in which he has been active for many years. He has written some of the fundamental papers in that field, including a 1961 paper entitled "Steps Toward Artificial Intelligence." To a large extent as a result of his interest in artificial intelligence, Minsky originally became interested in K. Eric Drexler's ideas about molecular nanotechnology. In the late 1980s, he chaired Drexler's doctoral committee and M.I.T., and he later wrote the introduction to Drexler's book *Engines of Creation*. Among Minsky's many awards have been the Turing Award of the Association for Computing Machinery (1970), the Japan Prize (1990), and the Rank Prize of the Royal Society of Medicine (1995).

Charles Musgrave (1966–)

Charles Musgrave was awarded the first Feynman Prize of the Foresight Institute in 1993 for his research on the use of a method for removing individual hydrogen atoms from a diamond surface. A report on this work, co-authored with Ralph Merkle, William Goddard, and J. K. Perry, appeared in the journal *Nanotechnology* in 1992.

Musgrave was born in Los Angeles on May 6, 1966. He earned his B.S. in materials science and engineering at the University of California at Berkeley. He then received an M.S. (1990) and a Ph.D. (1994) in materials science from the California Institute of Technology (Caltech). Musgrave spent a year as a postdoctoral student at Caltech and another year at the Massachusetts Institute of Technology. In 1995, he was appointed assistant professor of chemical engineering and materials science and engineering at Stanford University. Musgrave was Charles Powell Fellow at Stanford from 1996 to 1999 and is a member of Tau Beta Pi, Engineering Honor Society.

Christine L. Peterson (1957–)

Christine Peterson is a founding member of the Foresight Institute and its current executive director. She has been involved in molecular nanotechnology since it was first developed by her husband, K. Eric Drexler, in the 1980s.

Peterson was born in Buffalo, New York, on November 25, 1957. She received her B.S. from the Massachusetts Institute of Technology

(MIT) in 1979. She met Drexler at MIT and played a crucial role in encouraging and contributing to his development of the fundamental ideas of molecular nanotechnology. The two were married in 1981.

Peterson's role at the Foresight Institute has been primarily administrative and public education. She organizes the annual Foresight Conferences on Molecular Nanotechnology and is founding editor of the organization's *Foresight Updates*. She also is co-author (with Drexler and Gayle Pergamit) of the popular description of molecular nanotechnology, *Unbounding the Future* (William Morrow, 1991). Peterson describes her goal in lectures and written material as "making difficult technological concepts understandable and . . . lowering the stress of grappling with rapid technological change."

Calvin Quate (1923–)

Quate, Gerd Binnig, and Christoph Gerber collaborated on the invention of the atomic force microscope (AFM) in 1985. The AFM is a modification of the scanning tunneling microscope that makes possible new ways of manipulating matter at the atomic level.

Quate was born in Baker, Nevada, on December 7, 1923. He earned his B.S. from the University of Utah in 1944 and his Ph.D. from Stanford University in 1950. He worked at Bell Laboratories from 1949 to 1958 and at the Sandia Corporation from 1959 to 1961. In 1961, Quate joined the faculty of Stanford University, where he has remained ever since. He is currently Leland T. Edwards Professor of Engineering, professor of electrical engineering, and professor of applied physics. From 1983 to 1994, Quate also was senior research fellow at the Xerox Palo Alto Research Center.

Quate served on the Faculty of Sciences at Montpelier University, France, from 1968 to 1969. He was elected to membership in the National Academy of Engineering in 1970, the National Academy of Sciences in 1975, and the Royal Society of London in 1995. He was awarded the Rank Prize for OptoElectronics (1982), the IEEE Medal of Honor (1988), and the National Medal of Science (1992).

Mark Ratner (1942–)

Mark Ratner has made important contributions to the development of theoretical concepts regarding many aspects of chemical processes, including the development of molecular electronics. He was joint author (with Avi Aviram) of a seminal article on this topic in 1974.

Ratner was born in Cleveland, Ohio, on December 8, 1942. He received his A.B. from Harvard University in 1964 and his Ph.D. from Northwestern University in 1969. He did his postdoctoral studies at the Aarhus University in Denmark in 1970, then was an A. P. Sloan Foundation Fellow from 1972 to 1975. In 1975, Ratner accepted an appointment in the Department of Chemistry at Northwestern University, where he has remained ever since. He was department chair from 1990 to 1993 and associate dean in the College of Arts and Science from 1980 to 1984. Ranter has more than 375 professional publications and has written two textbooks, *Quantum Mechanics in Chemistry* (Prentice-Hall, 1993) and *Introduction to Quantum Mechanics in Chemistry* (Prentice-Hall, 2000).

Mark A. Reed (1955–)

Reed is one of the leading researchers in molecular electronics, the search for devices and methods at the nanoscale level that can be used to replace existing semiconductor-based devices and systems. He reported on the development of a molecular rectifier in 1997 and a molecular memory device in 1999.

Reed was born in Suffern, New York, on January 4, 1955. He received his B.S., M.S., and Ph.D. from Syracuse University in 1977, 1979, and 1983. After graduation from Syracuse, he accepted a position at the Central Research Laboratories of Texas Instruments. In 1990, he was appointed professor of electrical engineering and applied physics at Yale University, a post he has held ever since. He also is Harold Hodgkinson Professor of Engineering and Applied Science.

Among Reed's current research interests are quantum electron device physics; tunneling and transport phenomena in semiconductor heterojunction and nanostructured systems; resonant tunneling transistors, circuits, and devices; and quantum dot phenomena. Reed holds 12 U.S. and foreign patents and, with James Tour, is one of the founders of the Molecular Electronics Corporation.

Aristides A. G. Requicha (1939–)

Aristides A. G. Requicha is currently professor of computer science and electrical engineering at the University of Southern California and director of the Laboratory for Molecular Robotics. His primary research focus for the past 30 years has been the development of intelligent systems that interact with the three-dimensional world.

Requicha was born in Monte Estoril, Portugal, on March 18, 1939. He received his degree in electrical engineering from the Instituto Superior Técnico in Lisbon in 1962 and his Ph.D. in electrical engineering from the University of Rochester in 1970. He then accepted a position on the faculty at the University of Rochester, where he remained until 1986. He then moved to the University of Southern California, where he has served ever since. In addition to his current post, he is director of the university's Production Automation Project and has served as lecturer in physics at the University of Lisbon and as a research scientist with NATO's SACLANT Research Center in La Spezia, Italy.

Mihail C. Roco

Mihail C. Roco is currently chair of the National Science and Technology Council's committee on Nanoscale Science, Engineering, and Technology and Senior Advisor for Nanotechnology in the National Science Foundation. He played a critical role in the design and development of President Bill Clinton's National Nanotechnology Initiative.

Roco has received a number of honors, including the Carl Duisberg Award of the Federal Republic of Germany and recognition as Engineer of the Year in 1999, an award given by the National Society of Professional Engineers and the National Science Foundation. He is also a Corresponding Member of the Swiss Academy of Engineering Sciences and a Fellow of the American Society of Mechanical Engineers. He is currently Editor-in-Chief of the *Journal of Nanoparticle Research.*

Heinrich Rohrer (1933–)

Heinrich Rohrer and Gerd Binnig invented the scanning tunneling microscope (STM) in 1981 while they were employed at the IBM Zurich Research Laboratory. The STM has become one of the most powerful tools for the study of matter at the nanoscale level and an important device for the manipulation of individual atoms and molecules.

Rohrer was born in Buchs, St. Gallen, Switzerland on June 6, 1933. He attended the Swiss Federal Institute of Technology in Zurich, from which he received his Ph.D. in physics in 1960. Three years later, he joined the IBM Zurich Research Laboratory, where he has been employed ever since. Rohrer and Binnig were awarded a share of the 1986 Nobel Prize for Physics for their invention of the STM. In addition to the Nobel Prize, Rohrer has shared many awards with his colleague Binnig resulting from their work with the

STM. These awards include the German Physics Prize, the Otto Klung Prize, the Hewlett Packard Prize, and the King Faisal Prize.

Nadrian C. Seeman (1945–)

Nadrian C. Seeman is professor of chemistry at New York University. He was awarded the Foresight Institute's 1995 Feynman Prize in Nanotechnology for his research on synthetic DNA molecules with branching chains. Such molecules have the potential for use in the construction of complex molecular structures that will be needed in future molecular nanotechnology developments.

Seeman was born in Chicago on December 16, 1945. He was educated at the University of Chicago, from which he received his B.S. in biochemistry in 1966, and the University of Pittsburgh, from which he received his Ph.D. in crystallography in 1970. Seeman completed two postdoctoral programs, one at Columbia University in molecular graphics (1970–72) and one at the Massachusetts Institute of Technology in nucleic acid structure (1972–77).

Seeman's first teaching position was at the State University of New York at Albany. He moved to Columbia in 1987, where he has remained ever since. Seeman's special area of interest is in the use of DNA molecules to build precisely designed macromolecular crystals. For his work in this field, he was awarded the 1995 Feynman Prize. In addition to this award, Seeman has been honored with the Science and Technology Award from *Popular Science Magazine* (1993), the Emerging Technology Award from *Discover Magazine* (1997), and the Margaret and Herman Sokol Faculty Award in the Sciences (1999). He also was an honorary professor at the Universidad Peruana Cayetano Heredia in 1998.

Richard E. Smalley (1943–)

Richard E. Smalley was awarded a share of the 1997 Nobel Prize in Chemistry for his discovery of the 60-atom carbon isotope known as *buckminsterfullerene*. The isotope has proved to hold enormous potential as a building material from which nanoscale ropes, tubes, wires, and other devices can be constructed.

Smalley was born on June 6, 1943, in Akron, Ohio. He was educated at Hope College in Holland, Michigan, at the University of Michigan (B.S. in chemistry, 1965), and at Princeton University (M.A., 1971, and Ph.D., 1973). He did his postdoctoral studies at the James Franck Institute of the University of Chicago from 1973 to

1976. Smalley was offered a position in the Department of Chemistry at Rice University, where he since has served as assistant professor (1976–80), associate professor (1980–81), professor (1981–82), and Gene and Norman Hackerman Professor of Chemistry (1982– present). He also has been professor of physics at Rice since 1990.

Smalley has become one of the world's best-known spokespersons for research on the nanoscale level. In 1996, he was appointed director of Rice's new Center for Nanoscale Science and Technology. Among his many honors are the Irving Langmuir Prize in Chemical Physics of the American Physical Society (1991), the Welch Award in Chemistry of the Robert A. Welch Foundation (1992), the John Scott Award from the City of Philadelphia (1993), the Franklin Medal of the Committee on Science and the Arts of the Franklin Institute (1996), and the American Carbon Society Medal of the American Carbon Society (1997).

Deepak Srivastava (1957–)

Deepak Srivastava has authored or co-authored about 50 papers and given 100 invited and contributed presentations, including many on carbon nanotubes and nanotechnology. For more than 10 years, he has worked in a variety of research areas with fundamental significance to molecular nanotechnology, including the design and development of physical models, numerical methods, and software for the student of nanoscale materials and devices. He has chaired Foresight Conferences on Molecular Nanotechnology in 1998 and 1999, a Knowledge Foundation Conference on Carbon Nanotubes in 2000, and many sessions at other nanotechnology-related conferences throughout the United States. With the NASA Ames Computational Nanotechnology team, Srivastava is the co-winner of the Feynman Prize (theory) in molecular nanotechnology in 1997, NASA Ames Contractor Council Excellence Award in 1998, and a NASA Group Excellence Award in Devices and Nanotechnology in 2000.

Srivastava was born in Lucknow, India, on July 22, 1957. He earned his B.S. with honors at the University of Lucknow in 1977, an M.Ph. (experimental physics) from the University of Delhi in 1981, and a Ph.D. in theoretical physics at the University of Florida in 1988. Srivastava was a research assistant professor at Pennsylvania State University (1989–96) before joining the NASA Ames Research Center, where he is currently a senior scientist and task lead of the Computational Nanotechnology group.

James Tour (1959–)

James Tour has been a leading researcher in molecular electronics. He developed the first fully functioning molecular wire in 1996. The wire consisted of a modified polyphenylene molecule that has come to be known as a *Tour wire* in honor of its inventor.

Tour was born on August 18, 1959, in New York City. He received his B.S. in chemistry from Syracuse University in 1986 and his Ph.D. in organic chemistry from Purdue University in 1986. He worked as a postdoctoral student at the University of Wisconsin from 1986 to 1987 and at Stanford University from 1987 to 1988. He served as assistant professor, associate professor and professor in the Department of Chemistry and Biochemistry at the University of South Carolina from 1988 to 1996. He then was appointed to the faculty at Rice University in the Department of Chemistry and the Center for Nanoscale Science and Technology, a post he continues to hold. Tour's current areas of research interest include molecular electronics, molecular computing, self-assembly, self-replication, synthesis of molecular motors and nanotrucks, and regulation policies on precursors for chemical weapons of terror.

Steve Vetter (1957–)

Steve Vetter is president and chief executive officer of Molecular Manufacturing Enterprises, Inc., a seed capital firm formed to help accelerate advancements in the field of molecular nanotechnology. Vetter was born in Arlington Heights, Illinois, on May 16, 1957. He earned his B.S. and M.S. in software engineering at the University of Illinois in 1978 and 1980. He also pursued an M.B.A. degree at the University of Minnesota in 1984.

Vetter has been involved in the computer field for more than 20 years. He has founded or co-founded several new companies and written business plans for others. He created the Senior Associate program at the Foresight Institute and the Institute for Molecular Manufacturing. He also is senior advisor to the Space Studies Institute and was founder of the Minnesota Nanotechnology Special Interest Group.

James R. von Ehr, II (1950–)

James R. von Ehr, II, is the founder and chief executive officer of Zyvex LLC, the first nanotechnology development company. He earlier had founded the desktop publishing company, Altsys Corporation, which he sold in 1995 to Macromedia. He is a frequent speaker at industry trade shows and conferences.

Von Ehr was born in Grand Rapids, Michigan, on June 2, 1950. He received his B.S. in computer science from Michigan State University in 1973 and his M.S. in mathematical sciences at the University of Texas at Dallas in 1982. During the period from 1973 to 1984, von Ehr was employed at Texas Instruments, where he was manager in the Design Automation Department and senior member of the Technical Staff.

In 1984, von Ehr left Texas Instruments to found Altsys, a company devoted to the development of graphics applications for personal computers. Corporate sales eventually reached $100 million before von Ehr sold the company to Macromedia. From 1995 to 1997, von Ehr was an officer at Macromedia, serving as vice president and general manager of the Digital Arts Group and vice president of Product Development/Java. He founded Zyvex in 1997.

Von Ehr is author of many computer programs for graphics applications and computer games. He holds five patents in graphics applications.

John von Neumann (1903–57)

John von Neumann generally is regarded as one of the greatest mathematicians of the 20th century. His work covered a wide variety of important and creative subjects in economics, game theory, quantum physics, and computer design. In his book *Theory of Self-Reproducing Automata*, he presented one of the first sophisticated mathematical descriptions of machines that could produce other machines and that could make copies of themselves. These machines are, in some respects, early prototypes of Drexlerian assemblers and replicators.

von Neumann was born in Budapest, Hungary, on December 28, 1903. He studied mathematics at the University of Budapest and chemical engineering at the Eidgenössische Technische Hochschule in Zurich, from which he received his diploma in 1925. A year later, he completed his work for a Ph.D. in mathematics at the University of Budapest. He later did postdoctoral work at the University of Göttingen from 1926 to 1928.

von Neumann was appointed *privatdozent* (equivalent to assistant professor) at the University of Berlin in 1926. Three years later, he transferred to a similar post at the University of Hamburg. In 1930, von Neumann accepted a position at Princeton University, where he taught mathematics for 3 years. In 1933, he was appointed to a position in the Institute for Advanced Studies at Princeton, one of the most prestigious research institutions in the world. During World War II, von Neumann worked on the Manhattan Project, the program established by the U.S. government for the development of the first atomic bomb. von Neumann died in 1957 from bone cancer.

George M. Whitesides (1939–)

George M. Whitesides has long been involved in research with potentially significant relevance to the development of molecular nanotechnology. Among the topics in which he is interested are the fabrication of nanostructures and the processes of self-assembly. His views on molecular nanotechnology have drawn attention because he sees great promise for the development of nanoscale science and engineering but questions whether scientists will ever be able to develop some of the nanodevices described by Drexler in *Engines of Creation*. His interview with MIT's *Technology Review* in November/December 1998 is an excellent summary of his view (and that of other researchers) about the potential of molecular nanotechnology.

Whitesides was born August 3, 1939, in Louisville, Kentucky. He received his A.B. from Harvard University in 1960 and his Ph.D. from the California Institute of Technology in 1964. He was a member of the faculty of the Massachusetts Institute of Technology from 1963 to 1982. He joined the Department of Chemistry of Harvard University in 1982 and was department chair from 1986 through 1989. Whitesides is now Mallinckrodt Professor of Chemistry at Harvard University.

Whitesides has received many awards, including the American Chemical Society Award in Pure Chemistry in 1975, the James Flack Norris Award of the New England Section of the American Chemical Society in 1994, the Arthur C. Cope Award of the American Chemical Society in 1995, the Defense Advanced Research Projects Agency Award for Significant Technical Achievement in 1996, and the National Medal of Science in 1998.

Winners of the Feynman Prize in Molecular Nanotechnology

1993   Charles Musgrave, then a Ph.D. candidate in chemistry at the California Institute of Technology.

1995   Nadrian C. Seeman, Professor of Chemistry at New York University.

1997   Experimental
       A team of researchers from the IBM Research Division in Zurich and the
       CEMES-CNRS laboratory in France, consisting of James Gimzewski (IBM),
       Reto Schlittler (IBM), and Christian Joachim (CEMES-CNRS).

       Theoretical
       A team of researchers at the NASA Ames Research Center, consisting of
       Charles Bauschlicher, Stephen Barnard, Creon Levit, Glenn Deardorff, Al
       Globus, Jie Han, Richard Jaffe, Alessandra Ricca, Marzio Rosi, Deepak
       Srivastava, and H. Thuemmel.

1998   Experimental
       M. Reza Ghadiri, Scripps Research Institute

       Theoretical
       Ralph Merkle, then at Xerox Palo Alto Research Center, and Stephen Walch,
       ELORET, NASA Ames Research Center.

1999   Experimental
       Phaedon Avouris, IBM T. J. Watson Research Center in Yorktown Heights,
       New York.

       Theoretical
       A team of researchers at the California Institute of Technology, consisting of
       William Goddard, Tahir Cagin, and Yue Qi.

2000   Experimental
       R. Stanley Williams and Philip Kuekes, Hewlett Packard Laboratories, Palo
       Alto, California, and James Heath, University of California at Los Angeles.

       Theoretical
       Uzi Landman, Georgia Institute of Technology.

2001   Experimental
       Charles M. Lieber, Harvard University

       Theoretical
       Mark A. Ratner, Northwestern University

Chapter Five

Chronology

One of the most interesting ways to learn about nanotechnology is by reviewing important events that have taken place during the development of the science. This chapter lists some of those events with their significance for the current state of the field.

About 400 B.C.   The idea of an atom as a discrete, indestructible, fundamental particle of which all matter is composed first is proposed by the Greek philosopher Leucippus (*fl.* 430 B.C.) and his pupil Democritus (*ca.* 460–371 B.C.). The idea is developed further by the Roman philosopher Lucretius (*ca.* 95–55 B.C.) in his poem *De Rerum Natura*, then remains dormant for nearly 2 millennia.

1803   English chemist John Dalton (1766–1844) revives and clarifies the concept of an atom as the fundamental particle of which all matter is composed. The power of Dalton's atomic theory lay in his ability to provide a quantitative interpretation of the concept and to show how existing experimental data supported the existence of atoms. Dalton conceived of atoms as being hard, round, indestructible particles. This view of atoms remained dominant in chemical theory for about another century.

1897   English physicist J. J. Thomson (1856–1940) discovers the electron, showing that atoms are not solid and indestructible but instead consist of at least two charged parts, one the electron (negatively charged) and one a positive component as yet unidentified.

1926   Austrian physicist Erwin Schrödinger develops a series of partial differential equations that describe the properties of electrons in atoms by assuming that they exist as waves. From about this period onward, scientists tend to think about atoms less in terms of specific, concrete, physical objects and more in terms of mathematical equations (such as the Schrödinger wave equations) that explain observed behavior of atoms. The atom itself usually is pictured as a central nucleus surrounded by a "blob" of electrons, whose position is represented most accurately by a series of graphs showing the probability of finding electrons in various regions surrounding the nucleus.

1927 German physicist Werner Heisenberg (1901–1976) derives the Uncertainty Principle, which says that the position (x) and momentum (p) of a particle can never be determined simultaneously with an accuracy greater than a constant value known as *Planck's constant*. The principle has significant implications only for particles of very small mass, such as the electron. It means that scientists cannot be sure precisely where the electrons in an atom are at any particular moment in time.

1920s The theory of quantum mechanics is formulated by many different individuals, including Schrödinger, Heisenberg, Max Planck, Louis de Broglie, Albert Einstein, Max Born, and Paul Dirac. Arguably the most important implication of the theory is that matter has different properties at the submicroscopic or nanoscale level than it does at the macroscopic or microscopic level. The laws of physics and chemistry that apply to large masses of matter do not apply to individual particles, such as electrons.

1936 German physicist Erwin Wilhelm Müller (1911–1977) invents the field emission microscope (FEM). The FEM consists of a thin metal needle inside a vacuum tube. When strong current is applied to the needle, electrons are emitted from the tip and accelerated toward a fluorescent screen. The image displayed on the screen provides a likeness of the atomic structure of the metal that makes up the needle. Two decades later, Müller modifies the FEM to produce pictures of actual atoms (see *1953*).

1942 Robert Heinlein publishes a short story entitled "Waldo and the Magicians." The "Waldo" in the story is an inventor who creates a robot that can be controlled by human hands. Each movement of the human hands is replicated by the robot but by one half of its original dimensions. The robot, in turn, controls a second machine that replicates and reduces by half its own motions. This process is repeated over and over again until the movement of human hands is capable of producing tiny objects. The general principle behind the process described by Heinlein is incorporated later in Feynman's 1959 speech, "There's Plenty of Room at the Bottom."

1951 Hungarian-American mathematician John von Neumann publishes *The Theory of Self-Replicating Automata*, in which he analyzes the possibility of building machines that are able to duplicate themselves. The basic device suggested by von Neumann consists of two parts, a "universal computer," which can carry out any mathematical operation one wishes, and a "universal constructor," which can build any structure one wishes. von Neumann showed how the combination of these two machines could be used to make changes in the structure of a cell in two dimensions. Drexler's molecular replicators employ some of the principles developed by von Neumann extended to three dimensions.

1953   American biologist James Watson (1928–) and English biochemist Francis Crick (1916–) discover the molecular structure of deoxyribonucleic acid (DNA), the biochemical molecule that carries instructions for all chemical functions that take place within a cell. DNA and its cousins, the ribonucleic acids (RNAs), control the process by which individual atoms and groups of atoms are moved about and arranged inside cells so as to form large, complex molecules (proteins). Researchers in nanotechnology have found the DNA-RNA-protein model useful in exploring a variety of methods by which individual atoms and molecules can be used to build not only protein structures, but also other structures different from anything found in living cells.

1953   Erwin Müller (see *1936*) invents the field ion microscope (FIM), which provides clear images of individual atoms. The FIM is a modification of the FEM built by Müller in 1936 and accelerates ions (charged atoms) rather than electrons onto a fluorescent screen. Müller's photographs of individual atoms now are regarded as classic demonstrations that atoms do have specific and identifiable physical structures. One of his photographs of atoms was used on the cover of the June 1957 issue of *Scientific American* magazine.

1959   Richard Feynman presents a talk at the December meeting of the American Physical Society in Pasadena, California, in which he suggests the possibility of a new form of technology that involves the manipulation of individual atoms and molecules for the manufacture of new materials and structures. At the conclusion of his talk, Feynman offers two $1,000 prizes, one to the first person who could "take the information on the page of a book and put it on an area $1/25,000$ smaller in linear scale in such a manner that it can be read by an electron microscope" and one to the first person who can make an operating electrical motor "which can be controlled from the outside and, not counting the lead-in wires, is only $1/64$-inch cube." Feynman later repeats this talk on two other occasions, in 1983 and 1984.

1960   William H. McLellan, an engineer at Electro-Optical Systems in Pasadena, California, builds a motor that meets the conditions set by Feynman in his 1959 lecture. He collects his $1,000 check from Feynman.

1965   Gordon Moore, co-founder of the Intel Corporation, predicts that the number of circuits on a silicon chip would double every year. He later changed his projection to a doubling every 18 to 24 months. This prediction has since become known as *Moore's Law*.

1974   The possibility of constructing electrical rectifiers from single molecules or small molecular systems first is suggested by Ari Avram at the IBM Thomas J. Watson Research Center and Mark A. Ratner at Northwestern University.

1978   The biotechnology firm Genentech develops a method for producing human insulin artificially by using genetically engineered bacteria. A gene coding for the manufacture of human insulin is spliced into the genome of a strain of *E. coli* bacteria, which then begin to produce insulin spontaneously and automatically. The development is of significance to molecular nanotechnology because the methods used at Genentech are similar to those proposed for the construction of molecular assemblers.

1981   German physicists Gerd Binnig (1947–) and Heinrich Rohrer (1933–) invent the scanning tunneling microscope (STM), for which they eventually are awarded a share of the 1986 Nobel Prize for Physics. The invention is based on a property of electrons observable only at small dimensions, such as the nanometer scale. At these dimensions, electrons sometimes are able to flow across regions that would not be observed on the macroscopic or microscopic scales. Such behavior is known as *tunneling*. The STM is capable of making observations and measurements at the nanometer level and has been modified to be able to pick up and move individual atoms and molecules. It has the potential for becoming a powerful tool in the construction of nanoscale objects.

1981   The first scientific paper on molecular nanotechnology is published. The paper is K. Eric Drexler's "Molecular Engineering: An Approach to the Development of General Capabilities for Molecular Manipulation." It appears in the September 1981 issue of the *Proceedings of the National Academy of Sciences*. In the paper, Drexler lays out for the first time a formal outline of his view of molecular nanotechnology.

1985   Tony Newman, then a graduate student at Stanford University, develops a system for transcribing text at the electron-microscope level and earns the second prize promised by Richard Feynman in his 1959 lecture on "There's Plenty of Room at the Bottom." In the text written by Newman's system, each letter is about 60 atoms wide. With this technology, the whole of the *Encyclopedia Britannica* could be written on a surface 2 mm^2.

1985   A new allotrope of carbon is discovered by Richard Smalley, Robert Curl, and Harold Kroto. The allotrope consists of molecules of carbon containing 60 atoms each arranged in a nearly spherical structure consisting of hexagons and pentagons. The molecule is named *buckminsterfullerene* (or, more familiarly, a *buckyball*) in honor of the American architect Buckminster Fuller, who championed the use of this geometric shape in the construction of buildings. Scientists soon discover that buckyballs can be combined with each other so as to form long tubes known as *nanotubes*. Nanotubes have some interesting properties that make them ideal as construction units from which nanoscale devices can be constructed.

1985   Gerd Binnig, Christoph Gerber, and Calvin Quate invent the atomic force microscope (AFM). The AFM is a modification of the STM that Binnig and Heinrich Rohrer had invented 4 years earlier. It differs from the STM in that the microscope tip comes into contact with the surface it is studying, rather than being suspended a nanometer or so above the surface. Similar to the STM, the AFM has significant application to the field of nanotechnology.

1986   The first popular exposition on molecular nanotechnology, *Engines of Creation*, by K. Eric Drexler, is published.

1986   Drexler, Chris Peterson, and colleagues form the Foresight Institute. The purpose of the Institute is not primarily to promote research in the area of molecular nanotechnology, but to provide information about the technology and its potential consequences. The slogan that appears on the Institute's web page masthead is "Preparing for Nanotechnology."

1987   The Japanese government announces a Human Frontier Science Program in which nanotechnology is to play an important role. The program is planned to cover a 20-year period and to cost about $6 billion. According to a review by the Foresight Institute, "the understanding, design, and synthesis of molecular machines stands out as a major theme" of the program. In 1988, the program is re-evaluated, its goals clarified, and its budget reduced significantly. The United States and members of the European Community are invited to participate in the program.

1988   The first formal course in nanotechnology is offered at Stanford University. Drexler is the teacher. Topics include physical principles of molecular machines; the nature and methodology of exploratory engineering; nanocomputers, nanorobotics, and molecular assemblers; and applications of nanotechnology to computation, medicine, and large-scale systems.

1988   Scientists at the E. I. du Pont de Nemours Company, under the direction of William DeGrado, design and build the first completely artificial protein that spontaneously folds into a tertiary structure in water solution. The achievement is important to molecular nanotechnology because it shows the feasibility of a method for assembling atoms and groups of atoms into a structure that has some specific useful function.

1989   The First Foresight Conference on Nanotechnology is sponsored by the Foresight Institute and Global Business Network and hosted by the Stanford University Department of Computer Science. Reports presented at the meeting are later included in *Nanotechnology: Research and Perspective* (MIT Press, 1992).

1989   The University of Texas publishes *Assessing Molecular and Atomic Scale Technologies*, popularly known as the *MAST report*. The report summarizes research conducted by a group of graduate students at the University of Texas at Austin taught by Drexler. It outlines the views of many prominent scientists on the subject of molecular nanotechnology, lists some possible roadblocks to the development of the field, and makes recommendations about future directions for the science. The graduate students conclude that the most serious roadblock to the development of molecular nanotechnology may be its interdisciplinary nature, creating difficulties for researchers in different fields to communicate with each other on the subject.

1989   NanoCon, a 3-day conference on nanotechnology, is held in Seattle. Conference organizers list three major topics to be covered in the conference: (1) proposals for molecular design and engineering, (2) methods for the management of complex information systems, and (3) potential effects of developments in molecular nanotechnology over the coming 30 to 50 years. NanoCom proceedings are archived at http://www.halcyon.com/nano-jbl/NanoConProc/nanocon.html.

1989   Donald M. Eigler and Erhard K. Schweizer of the IBM Almaden Research Center spell out "IBM" by moving xenon atoms around on the surface of a nickel crystal by picking up individual atoms with an STM and placing them in position to spell out the three letters. The letter "I" requires 9 xenon atoms; the letter "B," 13; and the letter "M," 13. The exercise shows the feasibility of manipulating individual atoms with precision using the STM.

1990   British chemist J. Fraser Stoddart assembles a molecular-size railroad car consisting of about 24 atoms. He is able to move the car along a predetermined "track" and have it stop at any one of four distinct molecular "stations." As with the IBM demonstration cited previously, Stoddart's research has no specific practical objective, but it does show the feasibility of manipulating individual atoms in predetermined ways.

1990   Julius Rebek and colleagues in the Department of Chemistry at the Massachusetts Institute of Technology (MIT) design and construct the first artificial self-replicating molecule. The molecule splits apart, finds matching partners for each half, and builds two new molecules identical to the original molecule. The process then repeats itself endlessly as long as raw materials are available for the replication process. Rebek's system is similar to the process by which living cells reproduce by the process of mitosis.

1990   The English Institute of Physics launches a new journal called *Nanotechnology*. The journal is now a joint project of the English Institute of Physics and the American Institute of Physics.

1990   NANO I, the First International Conference on Nanometer Scale Science and Technology, is held in conjunction with STM '90, the Fifth International Conference on Scanning Tunneling Microscopy/Spectroscopy. The conference is sponsored jointly by the American Vacuum Society (AVS) and the U.S. Office of Naval Research. The nanotechnology segment of the conference becomes a regular and integral part of future AVS conferences.

1991   Drexler is awarded his Ph.D. in molecular nanotechnology by MIT. The degree is awarded by the MIT Media Lab, an interdisciplinary department that agreed to allow Drexler to study for his degree under its jurisdiction. The degree is the first Ph.D. in molecular nanotechnology ever awarded.

1991   Donald Eigler and colleagues at the IBM Almaden Research Center construct an "atomic switch" using an STM. The tip of the STM is placed a few nanometers above the surface of a layer of xenon atoms. When the voltage in the STM is varied, a single xenon atom is observed to jump back and forth between the surface and the needle tip. The research shows the ability of constructing familiar electronic devices at the atomic scale.

1991   The Institute for Molecular Manufacturing is founded in Palo Alto, California, for the purpose of promoting research in molecular nanotechnology and providing seed grants to researchers working in the field.

1991   The Second Foresight Conference on Molecular Nanotechnology is held in Palo Alto, California, sponsored by the Foresight Institute, the Institute for Molecular Manufacturing, the Department of Materials Science at Stanford

University, and the Research Center for Advanced Science and Technology of Tokyo University.

1991   The Center for Constitutional Issues in Technology (CCIT) is founded as an offshoot of the Foresight Institute. CCIT's purpose is to promote discussion on public-policy issues and to attempt to influence the thinking of public-policy sphere."

1991   Professor Dean L. Taylor in the Sibley School of Mechanical and Aerospace Engineering at Cornell University offers the first university courses in nanotechnology.

1991   A special division of the American Vacuum Society, the Nanometer-scale Science and Technology Division, is created. One of the division's first activities is sponsorship of NANO 3, the Third International Conference on Nanometer-scale Science and Technology. Among the topics discussed at the conference are manipulation of individual atoms, nanoscale biology, instrumentation and sensors, novel materials and methods of nanofabrication, and innovations in proximal probe technology.

1991   Japanese electron microscopist Sumio Ijima discovers the presence of carbon nanotubes in soot produced during the vaporization of carbon in an electric arc.

1991   The National Science Foundation initiates a nanotechnology research program called "Nanoparticle Synthesis and Processing."

1992   The First General Conference on Nanotechnology: Development, Applications, and Opportunities, is held in Palo Alto, California. The conference hears speakers on a wide range of topics, including computational molecular nanotechnology, biotechnology as an enabling technology, single-atom studies, growth of diamonoid materials, space applications, sponsorship of research in nanotechnology, nanotechnology in Japan, and the political context of research in molecular nanotechnology. Proceedings of the conference are reported in the book *Prospects in Nanotechnology: Toward Molecular Manufacturing* (John Wiley & Sons, 1995).

1992   Hearings on *New Technologies for a Sustainable World* are held in Washington, D.C., before the Subcommittee on Science, Technology, and Space of the U.S. Senate Committee on Commerce, Science, and Transportation (S.Hrg. 102–967). Drexler testifies before the committee about nanotechnology with the potential it may have for relieving poverty, disease, pollution, and other human problems. Chair of the subcommittee was Senator Al Gore, later to become Vice President under Bill Clinton (1992–2000).

1992   Researchers at the University of California at Irvine invent new kinds of ribonucleic acids capable of recognizing and manipulating an amino acid not found in natural proteins. The work illustrates that synthetic proteins can be made that will function as nanomachines.

1993   The Third Foresight Conference on Molecular Technology is held in Palo Alto, California, with its major focus on the use of computer-aided design of molecular systems.

1993   Scientists at the University of California at Berkeley invent a group of "artificial proteins" known as *oligocarbamates*. These substances have many structural similarities to natural proteins, but they are much stiffer. They hold promise for being useful as building units for nanoscale structures.

1993   The National Science Foundation agrees to fund the establishment and operation of a National Nanofabrication Users Network (NNUN) to carry out research on nanoscale devices. NNUN is to consist of five universities (Cornell, Howard, Pennsylvania State, Stanford, and the University of California at Santa Barbara) with equipment specially designed for such research. Scientists from other parts of the United States and other parts of the world are invited to apply to NNUN for research time at any one of the university sites.

1993   The first Feynman Prize in Nanotechnology is awarded to Charles Musgrave, a Ph.D. candidate in chemistry at the California Institute of Technology. The award is given for Musgrave's research on the modeling of a tool with which to remove hydrogen atoms from a surface. The Feynman Prize was established by the Foresight Institute to recognize "the researchers whose recent work has most advanced the development of molecular nanotechnology." The prize includes a cash award of $5,000.

1993   Rice University announces the establishment of a research and teaching program in the field of nanotechnology. The program will coordinate the efforts of faculty from the departments of chemistry, physics, biochemistry, mechanical engineering and materials science, electrical engineering, and chemical engineering. Richard Smalley, Nobel Prize winner for his discovery of buckyballs, is appointed director of the program.

1993   The Japanese government announces a major project designed to develop methods for the manipulation of matter at the molecular level. The project, called "Atom Technology," is planned to last 10 years and is awarded an initial grant of ¥25 billion (about $200 million). The program involves a consortium of 46 companies, 39 of which are Japanese. Consortium members include Hitachi, Toshiba, NEC, Fujitsu, Nippon Steel, Sumitomo Electric, Texas Instruments, Motorola, Dupont Japan, and Information Processing Technological Institute of Germany.

1993   Scientists first learn how to make single-wall carbon nanotubes (SWNTs). Previously the only kinds of carbon nanotubes available were multiwalled, of the type first discovered by Ijima in 1991. The availability of SWNTs makes it possible to determine the physical properties of carbon nanotubes with much greater precision.

1994   The first regular course in nanotechnology at the University of Southern California is offered by Professor Ari Requicha. The course is offered as a graduate course in the Department of Computer Sciences.

1994   Australian Prime Minister Paul Keating announces the government's inten-
        tion to spend $84 million (U.S. dollars) on seven new research centers, one
        of which will be a Nanotechnology Facility. The center will carry out research
        on nanotechnology devices designed for health care, food, and the environ-
        ment.

1994   A "Micro- and Nano-Engineering '94" conference is held in Davos, Switzer-
        land. The conference is important because it is the first year in the meeting's
        20-year history that nanotechnology is included in what previously has been
        an exclusively "microengineering" conference. New sessions included in the
        conference include "Atomic and Nanoscale Engineering" and "Nanoscale
        Fabrication and Devices."

1996   The Foresight Institute announces a $250,000 Feynman Grand Prize for
        major advances in molecular nanotechnology. The requirements for the prize
        are that an individual or group construct "a functional nanometer-scale ro-
        botic arm with specified performance characteristics, and also must design and
        construct a functional nanometer-scale computing device capable of adding
        two 8-bit binary numbers."

1996   NanoTech, a subsidiary of the Danish company BioSoft, holds what is said to
        be the first European conference on nanotechnology. The purpose of the
        conference, according to a company spokesman, is to "create a new Euro-
        pean Nanotechnology Initiative." Among the topics included in the confer-
        ence are supramolecular chemistry, self-assembling mechanisms and technol-
        ogy, computational chemistry and molecular modeling, materials science,
        engineering, and application areas.

1996   The International Business Communications conference on "Biological Ap-
        proaches and Novel Applications for Molecular Nanotechnology" is held in
        San Diego, California. The conference focuses on fabrication, characteriza-
        tion, outside world connections, and near-term applications of nanoscale de-
        vices.

1996   The Foresight Institute announces the award of a $1,000 prize to John M.
        Michelsen, a University of California at Irvine chemistry student for his work
        on "Atomically Precise, 3D Organic Nanofabrication: Reactive Lattice Sub-
        unit Design for Inverse AFM/STM Positioning." A year later, this award is
        formalized when the institute announces that it will make an annual award of
        $1,000 to the leading student in nanotechnology research.

1996   The Nobel Prize in Chemistry is awarded to R. F. Curl, H. W. Kroto, and
        R. E. Smalley for their research on the newly discovered allotrope of carbon
        known as *buckminsterfullerene*, a 60-carbon molecule. This molecule and its
        cousins, known collectively as *fullerenes*, are to become a fundamental build-
        ing block from which nanoscale objects are likely to be built.

1996   Staff writer Gary Stix writes a criticism of Drexlerian nanotechnology in *Sci-
        entific American* and sets off a storm of debate about the plausibility of
        Drexler's ideas.

1996   A team of researchers at the IBM Zurich Research Laboratory and France's National Center for Scientific Research show the ability to manipulate individual atoms and molecules with precision at room temperatures.

1996   A research team headed by James Tour, then of the University of South Carolina, and David Allara and Paul Weiss report on the invention of the first fully functioning molecular wire.

1996   Researchers at the IBM Research Division in Zurich announce the construction of a molecular-size abacus using 60-carbon molecules on a grooved copper surface.

1997   Robert M. Metzger at the University of Alabama at Tuscaloosa and Mark Reed at Yale University construct the first single-molecule rectifiers.

1997   *Discover Magazine* presents its Editor's Choice Award for Emerging Technology to an IBM research team from Zurich, headed by Dr. James Gimzewski, for their development of a "molecular abacus." The abacus consists of individual molecules as beads for counting. The beads are manipulated with a "finger" that consists of an ultrafine tip of an STM.

1997   California Molecular Electronics Corporation is formed by academic researchers and their business partners to conduct research and development projects for the commercialization of advances in the field of molecular electronics.

1997   The Nobel Prize in Chemistry is awarded to American chemist Paul Boyer and British chemist John E. Walker for their elucidation of the structure and function of "the smallest biological motor" yet discovered, the enzyme known as *ATPsynthase.*

1998   The European Commission announces that it will sponsor a series of three annual conferences in 1998, 1999, and 2000 on "Nanoscience for Nanotechnologies." One goal of the conferences is to create ongoing work groups on certain general topics in the field of nanotechnology. The conferences are open to members of the European Union and Iceland, Liechtenstein, Norway, and Israel.

1998   The National Institute of Standards and Technology sponsors a meeting in Albuquerque to define a new Advanced Technology Program intended to focus on microsystems and nanosystems. The topics on which the meeting focuses are commercialization, packaging, nanosystems, infrastructure, optics, devices, and materials.

1998   The National Science Foundation announces its first grant program in nanotechnology for fiscal year 1998. The program is called "Partnership in Nanotechnology: Synthesis, Processing, and Utilization of Functional Nanostructures" and provides a total of $10 million in grants.

1998   The Spring 1998 national meeting of the American Chemical Society includes a day-long session on "Device Applications of Nanoscale Materials" believed to be the first major symposium on the topic held at an American Chemical Society national meeting.

1998   A conference on "Nanotechnology for the Soldier System" is held in Cambridge, Massachusetts, to explore ways in which the new area of research has relevance to U.S. military programs. The conference is sponsored by the U.S. Army Soldier Systems Command, the Army Research Office, the Army Research Laboratory, and the National Science Foundation.

1998   The German government announces the establishment of a DM150 million ($90 million) nanotechnology research program to be located at six research and development centers in the country. The program is designed to focus on ultrathin functional layers, optoelectronic devices, development of lateral nanostructures, chemical functionalization of nanostructures, ultraprecise measurements, and nanostructure analysis.

1998   Researchers at IBM's Zurich Research Laboratory, the French National Center for Scientific Research in Toulouse, and the Riso National Laboratory in Roskilde, Denmark, announce the discovery of a molecular device that behaves as a nanomotor without the input of energy from human sources.

1998   The White House National Science and Technology Council creates the Interagency Working Group on Nanoscience, Engineering, and Technology, consisting of representatives from eight federal agencies, including the Departments of Commerce, Defense, and Energy; National Aeronautics and Space Administration; National Institutes of Health; National Science Foundation, White House Office of Science and Technology Policy of the National Science and Technology Council; and Office of Management and Budget.

1999   The Italian institute Electronics Biotechnology Advanced sponsors a conference in Rome in which molecular nanotechnology is a major focus of attention.

1999   The National Science Foundation sponsors a 3-day workshop on "Vision for Nanotechnology R&D in the Next Decade." A major goal of the meeting is the development of plans for a national nanotechnology initiative, to be included in President Clinton's fiscal year 2001 budget. The conference is attended by about 100 representatives from the National Science Foundation, National Aeronautics and Space Administration, National Institutes of Health, and Departments of Defense and Energy.

1999   Two winners in the 1999 Science Talent Search for high school students report on nanotechnology-related projects. Second Place winner David C. Moore, of Potomac, Maryland, did his research on ultrasmall electronic switches made from molecules. Tenth Place winner Alexander Wissner-Gross, of New Hyde Park, New York, simulated the use of buckyballs in the manufacture of nanoscale-electronic circuits.

1999   Researchers at the University of California at Santa Cruz and the Lawrence Berkeley National Laboratory report the complete structure of a bacterial ribosome at a resolution of 7.8 Å (0.78 nm). This work provides the most precise and complete picture of a ribosome yet available. Because of the ribosome's role in living systems as a bionanomachine, this accomplishment is of special interest and importance to those interested in molecular nanotechnology.

1999   Researchers at the Johnson Research Foundation at the University of Pennsylvania report on the construction of the largest synthetic protein ever made.

1999   Nathan Seeman and colleagues at New York University invent a robotic arm made of synthetic DNA stiff enough, yet flexible enough, to move molecules from one position to another.

1999   Researchers at the University of California at Los Angeles and the Hewlett-Packard corporation construct the first molecular switch.

1999   Mark Reed and colleagues at Yale University report success in constructing a primitive molecular memory device.

1999   Scientists at Northwestern University announce the invention of a nanopen, which draws lines only a few nanometers in width.

1999   Two reports of synthetic nanomotor devices are published in the September 9 issue of *Nature* magazine.

2000   The National Institutes of Health Office of Extramural Research sponsors a conference on "Nanoscience and Nanotechnology: Shaping Biomedical Research," in Bethesda, Maryland. The conference is attended by biomedical scientists, clinicians, and nanoscientists from physics, chemistry, computational, mathematics, and engineering fields. The conference announcement explains that the session will focus on "a new field of science and technology . . . that is not merely evolutionary but promises revolution in many aspects of our lives. Nanotechnology is emerging as a new field that has the potential to transform health care and medicine, biotechnology, manufacturing, energy, and information processing."

2000   The National Science and Technology Council published a 140-page report on the National Nanotechnology Initiative (NNI) that lays out President Clinton's 5-year vision for the federal government's role in advancing research and development in all fields of nanotechnology, including molecular nanotechnology. The President calls for a total of $495 million for the NNI. Although Congress allots only $423 million, that amount represents a $153 million (57%) increase over the previous year's budget.

2000   California Governor Gray Davis announces that the state will provide $25 million a year over a 4-year period to the California NanoSystems Institute. The Institute plans to raise $2 in private money for each $1 of state funding, making it the second largest nanotechnology program (after the federal NNI effort) in the United States.

2000   Researchers at the University of California at Berkeley report on the manufacture of a telescoping bearing in which an inner cylindrical core is capable of sliding in and out of and rotating within an outer shell with no apparent damage to either component.

2000   The Foresight Institute announces publication of the first public version of the "Foresight Guidelines on Molecular Nanotechnology." The purpose of the Guidelines is to ensure that research in the field proceeds in a safe and responsible manner. The Institute encourages researchers in the field to read and respond to the Guidelines and to design their research in accordance with the principles enunciated in the Guidelines.

2000   Researchers in molecular nanotechnology form the Molecular Electronics Corporation (MEC) to pursue commercial applications of their research. MEC is the second private corporation formed to promote research in molecular electronics, the first being California Molecular Electronics Corporation, formed in 1997.

2000   The American Chemical Society Division for Industrial and Engineering Chemistry announces the formation of a new technical subdivision for Advanced Materials and Nanotechnology. Some of the research areas to be included in the subdivision include carbon nanoscience (including carbon nanotubes and fullerenes), molecular-scale electronic devices, molecular self-assembly, and molecular-scale biomedical engineering.

2000   Flinders University, in Adelaide, South Australia, announced that it will be offering a Bachelor of Science degree in Nanotechnology. The program is described in detail at the University's website: http://adminwww.flinders.edu.au/Courses/ugrad/bachelor/nanotech.htm.

2000+   A number of universities announce the establishment of, or plans for, the creation of research centers on nanoscience and/or nanotechnology. Some of these institutions are Duke University (2001), Purdue University (2001), Tel Aviv University (2001), University of Albany (2001), University of Texas at Dallas (2001), and the Hebrew University in Jerusalem (2002).

2000+   A trend toward the creation of consortia for research in nanoscience and nanotechnology begins to emerge. These consortia typically consist of some combination of academic institutions, industrial corporations, governmental agencies, and private groups, often with the goal of dealing with all aspects of nanoscience and nanotechnology, from theoretical research to production of final product. Some examples of such consortia are the Ben Franklin Technology Partners of Southeastern Pennsylvania (Drexel University, University of Pennsylvania, and other groups, 2000), California NanoSystems Institute (University of California at Los Angeles and University of California at Santa Barbara, 2000), the I² NanoTech Centre in the United Kingdom (Queen Mary College of Birmingham University, QinetiQ, CLRC, AstraZeneca, GSK, and DTI Corporations, 2000), New Mexico Nanoscience Alliance (Los Alamos and Sandia National Laboratories and the University of New Mexico, 2001), and the Joint Institute for Nanoscience and Nanotechnology (University of Washington and the Department of Energy's Pacific Northwest National Laboratory, 2001).

2000+   A number of foreign governments announce an increased interest in research on nanoscience, often including the establishment of new research centers in this area. Some examples include the Nanotechnology Research Center in Beijing, China (2000), National Institute for Nanotechnology at the University of Alberta, Canada (2001), Minatec, a research center for microtechnology and nanotechnology at Grenoble, France (2001), a ten-year master plant to nurture nanotechnology in South Korea (2001), and the Center for Applied Nanotechnology to be located at the Industrial Technology Resource Institute in Hsin Chu, Taiwan (2002).

2001   The National Science Foundation announces funding for six centers for nanoscale research, with a total budget of $65 million. The centers are located at Columbia, Cornell, Harvard, Northwestern, and Rice universities, and at Rensselaer Polytechnic Institute.

2001   The U.S. Army announces plans for the establishment of an Institute for Soldier Nanotechnologies. The purpose of the institute is to develop nanometer-scale devices and materials that can be incorporated into a soldier's regular uniform and gear for such purposes as monitoring his or her life signs, providing camouflage or protection against weapons, and sensing the environment for toxic or hazardous materials.

2001   The National Science Foundation sponsors a workshop on Societal Implications of Nanoscience and Nanotechnology. The meeting focuses on the variety of ways in which developments in nanoscale research may affect social, political, economic, educational, and other aspects of the general culture.

Chapter Six

Documents

Sometimes the best way of understanding what is going on in a field of scientific research is to read what scientists and nonscientists are saying about the field. This chapter contains some important documents in the field of molecular nanotechnology, ranging from seminal manuscripts, such as Richard Feynman's original speech about "bottom-up" nanotechnology and Eric Drexler's first technical article on the topic, to more recent governmental and public statements about the status and potential applications of research in this field.

"There's Plenty of Room at the Bottom"

The following article is one of the handful of fundamental and classic papers in the area of nanotechnology. Written two decades before the concepts of nanotechnology were developed by any other research, Feynman provides a remarkable outline of what "bottom-up" might be able to do and how researchers might approach the task. He also speaks with amazing prescience about the problems and challenges of such research's becoming interdisciplinary in character.

But I am not afraid to consider the final question as to whether, ultimately—in the great future—we can arrange the atoms the way we want them; the very *atoms*, all the way down! What would happen if we could arrange the atoms one by one the way we want them (within reason, of course; you can't put them so that they are chemically unstable, for example).

Up to now, we have been content to dig in the ground to find minerals. We heat them and we do things on a large scale with them, and we hope to get a pure substance with just so much impurity, and so on. But we must always accept some atomic arrangement that nature gives us. We haven't got anything, say, with a "checkerboard" arrangement, with the impurity atoms exactly arranged 1000 angstroms apart, or in some other particular pattern.

What could we do with layered structures with just the right layers? What would the properties of materials be if we could really arrange the atoms the way we want them? They would be very interesting to investigate theoretically. I can't see exactly what would happen, but I can hardly doubt that when we have some control of the arrangement of things on a small scale we will get an enormously greater range of possible properties that substances can have, and of different things that we can do.

· · ·

ATOMS IN A SMALL WORLD

When we get to the very, very small world—say circuits of seven atoms—we have a lot of new things that would happen that represent completely new opportunities for design. Atoms on a small scale behave like *nothing* on a large scale, for they satisfy the laws of quantum mechanics. So, as we go down and fiddle around with the atoms down there, we are working with different laws, and we can expect to do different things. We can manufacture in different ways. We can use, not just circuits, but some system involving the quantized energy levels, or the interactions of quantized spins, etc.

Another thing we will notice is that, if we go down far enough, all of our devices can be mass produced so that they are absolutely perfect copies of one another. We cannot build two large machines so that the dimensions are exactly the same. But if your machine is only 100 atoms high, you only have to get it correct to one-half of one percent to make sure the other machine is exactly the same size—namely, 100 atoms high!

At the atomic level, we have new kinds of forces and new kinds of possibilities, new kinds of effects. The problems of manufacture and reproduction of materials will be quite different. I am, as I said, inspired by the biological phenomena in which chemical forces are used in repetitious fashion to produce all kinds of weird effects (one of which is the author).

The principles of physics, as far as I can see, do not speak against the possibility of maneuvering things atom by atom. It is not an attempt to violate any laws; it is something, in principle, that can be done; but in practice, it has not been done because we are too big.

Ultimately, we can do chemical synthesis. A chemist comes to us and says, "Look, I want a molecule that has the atoms arranged thus and so; make me that molecule." The chemist does a mysterious thing when he wants to make a molecule. He sees that it has got that ring, so he mixes this and that, and he shakes it, and he fiddles around. And, at the end of a difficult process, he usually does succeed in synthesizing what he wants. By the time I get my devices working, so that we can do it by physics, he will have figured out how to synthesize absolutely anything, so that this will really be useless.

But it is interesting that it would be, in principle, possible (I think) for a physicist to synthesize any chemical substance that the chemist writes down. Give the orders and the physicist synthesizes it. How? Put the atoms down where the chemist says, and so you make the substance. The problems of chemistry and biology can be greatly helped if our ability to see what we are doing, and to do the things on an atomic level, is ultimately developed—a development which I think cannot be avoided.

Now, you might say, "Who should do this and why should they do it?" Well, I pointed out a few of the economic applications, but I know that the reasons that you would do it might be just for fun. But have some fun! Let's have a competition between laboratories. Let one laboratory make a tiny motor which it sends to another lab which sends it back with a thing that fits inside the shaft of the first motor.

[Feynman then completes his article (and speech) with the two challenges for making nanoscale devices described in Chapter 1.]

Source: Richard Feynman, "There's Plenty of Room at the Bottom," *Engineering and Science*, February 1960, 22–36.

"Molecular Engineering: An Approach to the Development of General Capabilities for Molecular Manipulation"

This article can be considered as the first presentation of a technical argument for "bottom-up" nanotechnology. The article is remarkable in that it does not report the result of some research project, as is almost always the case in this journal. Instead, it outlines a research program and describes an unusual way of attacking a problem, an approach that has become a fundamental method of research in molecular nanotechnology. In the following selection, Drexler reminds the reader that he will be discussing general principles, without concern as to the specifics as to how those principles might be achieved or how they might be applied. That approach, he points out, falls within a long and honored scientific tradition, such as that which occurred in the evolution of the modern computer. End notes have been omitted from the passage extracted here.

Feynman's 1959 talk entitled "There's Plenty of Room at the Bottom" discussed microtechnology as a frontier to be pushed back, like the frontiers of high pressure, low temperature, or high vacuum. He suggested that ordinary machines could build smaller machines that would build still smaller machines, working step by step down toward the molecular level; he also suggested using particle beams to define two-dimensional patterns. Present microtechnology (exemplified by integrated circuits) has realized some of the potential outlined by Feynman by following the same basic approach: working down from the macroscopic scale to the microscopic.

Present microtechnology handles statistical populations of atoms. As the devices shrink, the atomic graininess of matter increasingly creates irregularities and imperfections, so long as atoms are handled in bulk, rather than individually. Indeed, such miniaturization of bulk processes seems unable to reach the ultimate level of microtechnology—the structuring of matter to complex atomic specifications. In this paper, I will outline a path to this goal, a general molecular engineering technology. The existence of this path will be shown to have implications for the present.

Although the capabilities described may not prove necessary to the achievement of any particular objective, they will prove sufficient for the achievement of an extraordinary range of objectives in which the structuring and analysis of matter are concerned. The claim that devices can be built to complex atomic specifications should not, however, be construed to deny the inevitability of a finite error rate arising from thermodynamic effects (and radiation damage). Such errors can be minimized through the use of free energy in error-correcting procedures (including rejection of faulty components before device assembly); the effects of errors can be minimized through fault-tolerant design, as in macroscopic engineering.

The emphasis on devices that have general capabilities should be taken in the spirit of early work on the theoretical capabilities of computers, which did not attempt to predict such practical embodiments as specialized or distributed computational systems. The present argument, however, will proceed from step to step by close analogies between the proposed steps and past developments in nature and technology, rather than by mathematical proof. We commonly accept the feasibility of new devices without formal proof, where analogies to existing systems are close enough: consider the feasibility of making a clock from zirconium. The detailed design of many specific devices to render them describable by dynamical equations would be a task of another order (consider designing a clock from scratch) and appears unnecessary to the establishment of the feasibility of certain general capabilities.

Source: K. Eric Drexler, "Molecular Engineering: An Approach to the Development of General Capabilities for Molecular Manipulation," *Proceedings of the National Academy of Sciences*, September 1981, 5275–78. Reprinted by permission of the author. This article also is available online at http://www.imm.org/PNAS.html.

"Nanotechnology: The State of Nano-Science and Its Prospects for the Next Decade"

On June 22, 1999, the Subcommittee on Basic Research of the U.S. House of Representatives held a hearing on nanotechnology. Experts who testified at the hearing were Dr. Eugene Wong, Assistant Director of the National Science Foundation's Engineering Department; Dr. Richard Smalley, of Rice University; Dr. Ralph C. Merkle, of XEROX; and Mr. Paul McWhorter, Deputy Director of Sandia National Laboratories' Microsystems Science, Technology and Components Center. As outlined by Subcommittee Chair Nick Smith (RMI), the purpose of the hearing was to "review federal funding of nanotechnology research, to discuss the role of the federal government in supporting nano-science research, and to discuss the economic implications of scientific advances made in the field of technology."

As introduction to the subject, the fiscal year 1999 federal funding of nanotechnology research was provided. That information is as follows:

Fiscal Year 1999 Federal Funding of Nanotechnology Research (Millions of Dollars)

Federal Agency	Fiscal Year 1999
National Science Foundation	80
Department of Defense	60
Department of Energy	54
National Institutes of Health	5
Department of Commerce	12
National Aeronautics and Space Administration	18
Total	232

A portion of the testimony provided by Merkle is abstracted. Note that in expert testimony, there may or may not have been some attempt to distinguish between "top-down" and "bottom-up" nanotechnology.

Introduction

For centuries manufacturing methods have gotten more precise, less expensive, and more flexible. In the next few decades, we will approach the limits of these trends. The limit of precision is the ability to get every atom where we want it. The limit of low cost is set by the cost of the raw materials and the energy involved in manufacture. The limit of flexibility is the ability to arrange atoms in all the patterns permitted by physical law.

Most scientists agree we will approach these limits, but differ about how best to proceed, on what nanotechnology will look like, and how long it will take to develop. Much of this disagreement is caused by the simple fact that, collectively, we have only recently agreed that the goal is feasible and we have not yet sorted out the issues that this creates. This process of creating a greater shared understanding both of the goals of nanotechnology and the routes for achieving those goals is the most important result of today's research.

THE GOAL

Nanotechnology (or molecular nanotechnology to refer more specifically to the goals discussed here) will let us continue the historical trends in manufacturing right up to the fundamental limits imposed by physical law. It will let us make remarkably powerful molecular computers. It will let us make materials over fifty times lighter than steel or aluminum alloy but with the same strength. We'll be able to make jets, rockets, cars or even chairs that, by today's standards, would be remarkably light, strong, and inexpensive. Molecular surgical tools, guided by molecular computers and injected into the blood stream could find and destroy cancer cells or invading bacteria, unclog arteries, or provide oxygen when the circulation is impaired.

Nanotechnology will replace our entire manufacturing base with a new, radically more precise, radically less expensive, and radically more flexible way of making products. The aim is not simply to replace today's computer chip making plants, but also to replace the assembly lines for cars, televisions, telephones, books, surgical tools, missiles, bookcases, airplanes, tractors, and all the rest. The objective is a pervasive change in manufacturing, a change that will leave virtually no product untouched. Economic progress and military readiness in the 21st Century will depend fundamentally on maintaining a competitive position in nanotechnology.

SELF REPLICATION AND LOW COST

Many researchers think self replication will be the key to unlocking nanotechnologies full potential [sic], moving it from a laboratory curiosity able to expensively make a few small molecular machines and a handful of valuable products to a robust manufacturing technology able to make myriads of products for the whole planet. We know self replication can inexpensively make complex products with great precision: cells are programmed by DNA to replicate and made complex systems—including giant redwoods, wheat, whales, birds, pumpkins, and more. We should likewise be able to develop artificial programmable self replicating molecule machine systems—also known as assemblers—able to make a wide range of products from graphite, diamond, and other non-biological materials. The first groups to develop assemblers will have a historic window for economic, military, and environmental impact.

WHAT NEEDS TO BE DONE

Developing nanotechnology will be a major project—just as developing nuclear weapons or lunar rockets were major projects. We must first focus on developing new things: the tools with which to build the first molecular machines and the blueprints of what we are to build. This will require the cooperative efforts of researchers across a wide range of disciplines: scanning probe microscopy, supramolecular chemistry, protein engineering, self assembly, robotics, materials science, computational chemistry, self replicating systems, physics, computer science, and more. This work must focus on fundamentally new approaches and methods: incremental or evolutionary improvements will not be sufficient. Government funding is both appropriate and essential for several reasons: the benefits will be pervasive across companies and the economy; few if any companies will have the resources to pursue this alone; and development will take many years to a few decades (beyond the planning horizon of most private organizations).

We know it's possible. We know it's valuable. We should do it.

Source: Committee on Science, U.S. House of Representatives, Subcommittee on Basic Research, Hearings on Nanotechnology: The State of Nano-Science and Its Prospects for the Next Decade, 22 June 1999. Also available on the Internet at http://www. house.gov/science/basic_charter_062299.htm and at http://www.merkle.com/ nanohearing1999.html.

The following section is taken from the written statement offered by Smalley at the Subcommittee hearing. It consists of two parts, the first of which is Smalley's own statement. The second part of the statement is Smalley's abstraction of the Draft Executive Summary of a report by the Interagency Working Group on Nanoscience, Engineering, and Technology (IWGN), prepared for the Office of Science and Technology Policy Committee on Technology on March 10, 1999. The full report is available on the Internet at the website for this hearing.

Mr. Chairman, I appreciate the opportunity to present my views on nanotechnology. There is a growing sense in the scientific and technical community that we are about to enter a golden new era. We are about to be able to build things that work on the smallest possible length scales, atom by atom with the ultimate level of finesse. These little nanothings, and the technology that assembles and manipulates them—nanotechnology—will revolutionize our industries, and our lives.

Everything we see around us is made of atoms, the tiny elemental building blocks of matter. From stone to copper, to bronze, iron, steel, and now
silicon, the major technological ages of humankind have been defined by
what these atoms can do in huge aggregates, trillions upon trillions of atoms
at a time, molded, shaped, and refined as macroscopic objects. Even in our
vaunted microelectronics of 1999, in our highest-tech silicon computer chip
the smallest feature is a mountain compared to the size of a single atom.
The resultant technology of our 20th century is fantastic, but it pales when
compared to what will be possible when we learn to build things at the ultimate level of control, one atom at a time.

Nature has played the game at this level for billions of years, building stuff
with atomic precision. Every living thing is made of cells that are chock full of
nanomachines—proteins, DNA, RNA, etc.—each jiggling around in the
water of the cell, rubbing up against other molecules, going about the business of life. Each one is perfect right down to the last atom. The workings are
so exquisite that changing the location or identity of any atom would cause
damage. Over the past century, we have learned about the workings of these
biological nanomachines to an incredible level of detail, and the benefits of
this knowledge are beginning to be felt in medicine. In coming decades we
will learn to modify this machinery to extend the quality and length of life.
Biotechnology was the first nanotechnology, and it has a long way yet to go.

Let me give you just one, personal, example: cancer. I sit before you
today with very little hair on my head. It fell out a few weeks ago as a result
of the chemotherapy I've been undergoing to treat a type of non-Hodgkin's
lymphoma—the same sort that recently killed King Hussein of Jordan.
While I am very optimistic, this chemotherapy is a very blunt tool. It consists of small molecules which are toxic—they kill cells in my body. Although they are meant to kill only the cancer cells, they kill hair cells too,
and cause all sorts of other havoc.

Now, I'm not complaining. Twenty years ago, without even this crude
chemotherapy I would already be dead. But twenty years from now, I am
confident we will no longer have to use this blunt tool. By then nanotechnology will have given us specifically engineered drugs which are nanoscale
cancer-seeking missiles, a molecular technology that specifically targets just
the mutant cancer cells in the human body, and leaves everything else bliss-
fully alone. To do this these drug molecules will have to be big enough—
thousands of atoms—so that we can code the information into them of
where they should go and what they should kill. They will be examples of an
exquisite human-made nanotechnology of the future. I may not live to see
it. But, with your help, I am confident it will happen. Cancer—at least the
type that I have—will be a thing of the past.

Powerful as it will be, this bio-side of nanotechnology that works in the water-based world of living things will not be able to do everything. It cannot make things strong like steel or conduct electricity with the speed and efficiency of copper or silicon. For this, other nanotechnologies will be developed—what I call the "dry side" of nanotech. My own research these days is focused on carbon nanotubes—an outgrowth of the research that led to the Nobel Prize a few years ago. These nanotubes are incredible. They are expected to produce fibers 100 times stronger than steel at only $\frac{1}{6}$th the weight—almost certainly the strongest fibers that will ever be made out of anything. In addition they will conduct electricity better than copper. Membranes made from arrays of these nanotubes are expected to have revolutionary impact in the technology of rechargeable batteries and fuel cells, perhaps giving us all-electric vehicles within the next 10–20 years with the performance and range of a Corvette at a fraction of the cost.

As individual nanoscale molecules, carbon nanotubes are unique. They have been shown to be true molecular wires, and have already been assembled into the first molecular transistor ever built. Several decades from now we may see our current silicon-based microelectronics supplanted by a carbo-based nanoelectronics of vastly greater power and scope.

It's amazing what one can do just by putting atoms where you want them to go.

[Smalley then reads into the record a summary of the Interagency report.]

DRAFT—EXECUTIVE SUMMARY—DRAFT

Recommendation:

As part of the fiscal year 2001 budget, the IWGN recommends a national initiative. The initiative, known as NTR (Nanotechnology for the Twenty-First Century: Leading to a New Industrial Revolution), will approximately double the Federal Government's annual investment in nanotechnology research and development from its present (FY99) base of $234M per year. The increase will be incrementally grown over a three-year interval.

The NTR Initiative will address five activities:

Long-term nano science and engineering research that will lead to fundamental understanding and to discoveries of novel phenomena, processes and tools for nanotechnology. This commitment will refocus the government investment beginning in the 1950s that led to today's microelectronics, microfabrication, and computer technology;

Synthesis and processing "by design" of engineered, nanometer-size, material building blocks and system components, fully exploiting molecular self-assembly concepts. This commitment will generate new classes of high performance materials, bio-inspired systems, paradigm changes in device design, and efficient, affordable manufacturing of high performance products. Novel properties and phenomena will be enabled, as control of structures of atoms, molecules and clusters becomes possible;

Nanodevice concepts including system architecture research to best exploit nano-derived properties in operational systems, and combining building-up of molecular structures with ultraminiaturization. The new nanodevices will cause orders of magnitude improvements in microprocessors and mass storage, create tiny medical tools that minimize collateral damage; and enable uninhabited defense combat vehicles in fully imaged battle fields. There will be dramatic payback to programs with this National priority including information technology, nanobiotechnology and medical technology;

Application of nanostructured materials and systems to manufacturing, power systems, energy, environment, national security, and health. Areas of interest include advanced dispersions, catalysts, separation methods, and consolidated nanostructures, as well as increase the pace of knowledge and technology transfer;

Educate and train a new generation of skilled workers in the multidisciplinary perspectives necessary for rapid progress in nanotechnology.

Potential for NTR impact is compelling:

Nano science and engineering knowledge is exploding worldwide because of the availability of new investigative tools; maturity in the biology, chemistry, engineering, materials and physics disciplines, and interdisciplinary synergism; and financial support driven by emerging technologies and their markets. The science and engineering communities have generated a flurry of new results, doubling the publication rate each two–three years. During 1998, funding agency initiatives in nanotechnology (NSF Functional Nanostructures Initiative, and DoD Multidisciplinary University Research Initiative in Nanoscience) had success rates no higher than 1 in 6, constrained only by funding limitations;

The nanotechnology revolution will lead to fundamental breakthroughs in the way materials, devices and systems are understood, designed and manufactured. Dr. Neal Lane stated at a Congressional hearing in April 1998 that "If I were asked for an area of science and engineering that will most likely produce the breakthroughs of tomorrow, I would point to nanoscale science and engineering." Potential breakthroughs include orders-of-magnitude increases in computer efficiency; emergence of entirely new phenomena in physics and chemistry; nanofabrication of three-dimensional molecular architecture; novel processing architectures such as quantum computer and cellular automata; repair of human body with replacement parts; and a virtual presence in space;

Nanotechnology is creating a revolution in the way materials, devices and systems are manufactured and perform. In the last few years, applying fundamental discoveries has developed multi-billion dollar product lines. These include: giant magnetoresistance multilayers (for computer memory), nanostructured coatings (in the data storage and photographic industry), nanoparticles (colorants in printing and drug delivery in pharmaceutical field), superlattice confinement effects (for optoelectronic devices and lasers), and nanostructured materials (nanocomposites and nanophase metals). John Armstrong, formerly Chief Scientist of IBM, wrote in 1991, "I believe nanoscience and nanotechnology will be central to the next epoch of the information age, and will be as revolutionary as science and technology at the micron scale have been since the early '70s." More recently, industry leaders including those at the IWGN workshop on Jan. 27–29, 1999, have extended this vision by concluding that nanoscience and technology will change the nature of almost every human-made object in the next century.

Nanoscience is an opportunity to energize the interdisciplinary connections between biology, chemistry, engineering, materials, mathematics, and physics in education. It will give birth to new fields that are only envisioned at this moment;

European and Pacific countries have developed focussed programs in the science and technology of nanostructures that will provide world-wide critical mass to this initiative, accelerate progress, and guarantee commercial competition for the results.

NTR investment strategy:

This initiative builds on previous and current nanotechnology programs, including some early investment from the Advanced Materials Processing Program, NSF instrumentation and functional nanostructures, and DoD programs supporting its Nanoscience Strategic Research Objective;

The lead-time for science maturing into technology is approximately 10–15 years; now is a critical time for government investment in the S&T of nanostructures. The leaders from industry, academe and government present at the IWGN Workshop concluded that the Federal Government was underinvesting in long-term nanotechnology research and development relative to the outstanding opportunities. The private sector is unlikely to invest in nano science and engineering research until products are 3–5 years from commercialization;

Roughly 70 percent of the funding will be for university-based research, which will also help meet the demand for skilled workers with advanced nanotechnology skills in the next century. In the academic programs, it is anticipated that 65% of the funding will be for single investigators, 15% for multidisciplinary programs and 5–10% for nanotechnology centers that will play a similar role to the supercomputer centers, 5–10% instrumentation development and procurement, and 5% for the development of multidisciplinary educational programs.

Government/industry/academic partnerships will be strongly encouraged.

Source: Committee on Science, U.S. House of Representatives, Subcommittee on Basic Research, Hearings on Nanotechnology. "The State of Nano-Science and Its Prospects for the Next Decade," June 22, 1999. Also available on the Internet at http://www.house.gov/science/basic_charter_062299.htm

Federal Policy on Research in Nanotechnology (I)

The White House Office of Science and Technology Policy (OSTP) was created in 1976 to advise the President on scientific and technological issues. In 1995, then Director of the OSTP, Dr. Jack Gibbons, presented his views on the development of molecular nanotechnology to the National Conference on Manufacturing Needs of U.S. Industry held at the National Institute of Standards and Technology. His remarks on the subject were as follows:

Nanoscience has become an engineering practice. Based on recent theoretical and experimental advances in nanoscience and nanotechnology, precise atomic and molecular control in the synthesis of solid state three-dimensional nano-structures is now possible. The volume of such structures is about a billionth of that of structures on the micro scale.

The next step is the emergence of nanotechnology. The stage is being set, I believe, for actual manufacture of a wide variety and range of custom-made products based on the ability to manipulate individual atoms and molecules during the manufacturing process. The ability to synthesize devices such as molecular wires, resistors, diodes, and photosynthesis elements to be inserted in nanoscale machines is now emerging from fundamental nanoscience. Already the use of optical materials assembled at the molecular level has revolutionized response time, energy losses, and transport efficiency in nanoscale materials.

Next, molecular manufacturing for mass production of miniature switches or valves or motors or accelerometers, all at affordable prices, is a genuine possibility in the not so distant future. This new technology could fuel a powerful economic engine providing new sources of jobs and wealth and technology spillovers.

Further fundamental understanding of basic physical phenomena at the quantum level will be needed to understand and reach these kinds of technological opportunities. Some of the areas in which knowledge must be deepened are superlattices and multiquantum wells, localization effects of electron and light waves, flux patterns and their pinning, and dynamics in superconductors, as well as further quantum mechanical analysis of nanostructured systems. This basic scientific understanding will find a very broad range of technological applications, from energy storage and generation to magnetic storage and recording, to supercomputers.

To an ex-physicist like me, these prospects for scientific exploration are exhilarating, and our new understanding of a complex, symbiotic relation between science and technology—rather than a simple hand-off—makes the prospects still even more exciting. But my post-physics years of starting with new high technology companies beyond physics and then doing policy work at the Office of Technology Assessment, and my present deep immersion in policy at the White House Office of Science and Technology Policy, remind me that the reduction of leading-edge technologies to practice is a process which, as you so full well know, can be risky and arduous. It's a long, long way from invention to profitable production.

Cooperative efforts by government and industry to advance technology can help fill that gap. One of this Administration's top priorities is to form closer working partnerships with industry, as well as with universities, state and local governments, and workers, to strengthen America's industrial competitiveness and create jobs.

Source: Speech presented at the National Conference on Manufacturing Needs of U.S. Industry, National Institute of Standards and Technology, March 1997.

Federal Policy on Research in Nanotechnology (II)

Research in the area of nanotechnology continued to be a topic of interest to Gibbons's successor as Director of the OSTP, Dr. Neal Lane. Dr. Lane spoke about his role in the development of federal policy on nanotechnology research at a symposium honoring President Chun Lin at the University of Oklahoma on October 13, 2000.

Twenty years ago, the tunneling/scanning microscope was invented at IBM. That marked the first time we could see a cluster of molecules on a surface. But to observe and identify was just the beginning. To learn to custom assemble microscopic structures and devices with unique properties and capabilities is the ultimate goal. We can anticipate everything from machines on the scale of human cells to auto tires designed atom by atom. This new work will require many fields of science and engineering working together.

Any research wave builds by the free and open disclosure of new knowledge. That sharing of knowledge, its replication by experiments, and the cross-communication of researchers in the field and beyond is the scientific process we all know.

These time-honored practices create vibrations in the research community that "something new" is happening. In 1996, when I was Director of NSF, I sensed a growing wave of interest in the research community to greatly expand research activities at the nanoscale. Work in this emerging field cut across several NSF Directorates so I asked one of our staff scientists to look into creating an informal working group on nanotechnology. I wanted a bottoms-up approach of talking to individual researchers at the program level because at that point nanotechnology was not seen as a priority. Support had to be built one researcher at a time during 1997.

Between 1996–1998, a group of federal agencies, under the auspices of NSF, commissioned a worldwide study. It would assess the current status and future trends in R&D, and especially in the growing areas of nanoscale science and technology. We wanted to determine the most effective strategy to invest Federal R&D funds to develop U.S. potential in the coming years. The study report was issued in December 1998. But even before that, NSF held a competition called Functional Nanostructures. Other agencies were invited to co-review and co-fund projects. The competition was over subscribed and almost 40 percent of the projects were funded by other agencies. The Department of Defense also announced a competition. They got a response far greater than anticipated. The air was thickening with anticipation and excitement. Also, in September 1998, the NSTC established an Interagency Working Group on Nanotechnology. This effort assured cross-agency cooperation and also raised the visibility of the issue to the presidential level. Plans emerged for a workshop to bring all nanoscale activities together under one umbrella.

By January 1999, a university, industry, government workshop on research priorities met. The basic vision of the National Nanotechnology Initiative was created. Here, the NSTC role was critical in building consensus and accelerating the timetable. But an important link was still missing. We asked PCAST to assemble a panel on nanotechnology to review the possible Nanotechnology Initiative. Charles Vest, President of MIT, chaired the group, which was comprised of five members from academe and five from industry. They provided the outside point of view. Their review resulted in some very constructive changes. In the end, we not only got a thumbs-up, but even the next rung, important enough for a Presidential Initiative. Don't get the idea that it was free sailing from here. Turning the vision into a Presidential Initiative required those critical skills that we so often underestimate—building trust and confidence, communicating clearly, making sure that all parties are on board.

The 11 months from March of 1999 to February 2000 were the most feverish and maybe the roughest. They included, among other things, Congressional hearings on nanotechnology in May and June. In February 2000, the initiative was announced including the supplement to the President's budget. This is perhaps more than you ever wanted to know but highly illustrative of the painstaking process of developing a Presidential Initiative from "something new" that bubbles up from the research community. It's a difficult road that has every imaginable pothole and pitfall. Although the process may sound drawn-out in the telling, we went from building individual researcher support in 1997 to the announcement of the full Initiative package in a tightly compressed 3-year period. This requires superb staff work on the part of so many people.

In finality, it represents a vision of a "big idea," an idea that required people with telescopes. And those people had to have the courage to stubbornly defend their vision to those who doubted they had the telescopes. Nanotechnology can become the next territory of enabling research, perhaps as pervasive as information technology continues to be. Other nations have also done important work in this field. Our goal in the Initiative is to nurture the field and support it in depth to expand the core knowledge and in breadth to pollinate its diverse application across many fields. But emerging fields always rise from the strength of established fields. So our challenge is to advance the newest frontier while keeping fundamental disciplines strong. It's a balancing act—and we have to get it right.

Source: Speech by Dr. Neal Lane President Chun Lin, Symposium, University of Oklahoma, October 13, 2000.

New Technologies for a Sustainable World

On June 26, 1992, then Senator Al Gore (D-Ky), later Vice President, convened a hearing on the ways in which new forms of technology could be used to contribute to a more advanced world with less environmental degradation. K. Eric Drexler, Chairman of the Foresight Institute, presented the following testimony before this hearing.

Mr. Chairman, I would like to thank you and the members of this subcommittee for this opportunity to discuss a topic that I expect will one day become a leading issue in these halls. The focus of this hearing—new technologies for a sustainable world—is particularly appropriate for discussion of this topic, because a concern with the consequences of future technologies for the environment and for the human condition has for many years guided my research, and has led to the results described here.

In the decade since I first described molecular nanotechnology in the *Proceedings of the National Academy of Sciences*, this field has progressed from general theoretical concepts to early laboratory demonstrations and a growing body of detailed designs. Five years ago, audiences questioned whether individual atoms could be placed in precise patterns; today, I can answer that question not just with calculations, but with a slide showing the letters "IBM" spelled using 35 xenon atoms.

The Foresight Institute, which I serve as Chairman, sponsors a series of scientific conferences on molecular nanotechnology. The most recent, held last autumn, was co-sponsored by the Stanford University Department of Materials Science and Engineering and the University of Tokyo Research Center for Advanced Science and Technology; this meeting has stimulated at least three laboratory research efforts directed toward a key milestone on the path to molecular nanotechnology. Japan's Ministry of International Trade and Industry recently committed some $185 million over the next ten years to a nanotechnology research effort; development of molecular systems is seen in Japan as fitting with the broad goal of developing environmentally-compatible technologies.

Momentum toward the development of molecular nanotechnology is building around the world. The consequences for human life and for Earth's environment will be enormous, and could be enormously positive. The balance of this testimony begins by describing molecular nanotechnology from a biological and ecological perspective and sketching some of its wide range of applications. It then describes the relevant areas of research; the level of activity in the U.S., Japan, and Europe; and some of the policy issues that its development can be expected to raise. The closing section discusses how these concepts can be evaluated before committing to any substantial effort that presumes their validity.

MOLECULAR NANOTECHNOLOGY: A BIOLOGICAL AND ECOLOGICAL PERSPECTIVE

Industry today consumes fossil fuel and discharges carbon dioxide into the atmosphere. Forests and farms, in contrast, produce useful products (including fuels) while removing carbon dioxide from the atmosphere. Proposals for reducing the concentration of greenhouse gases typically focus on modifying existing industrial technologies to reduce emissions, and this is a sound strategy. Yet it may be better to develop industrial technologies that, like forests and farms, are carbon dioxide consumers.

Leaves are solar energy collectors employing molecular electronic devices: chlorophyll molecules and photosynthetic reaction centers. These solar energy collectors, like the other useful products of forests and farms, are built by systems of molecular machinery such as ribosomes and metabolic enzymes. A natural direction for technology, then, is to learn to apply systems of molecular machinery to build useful products in industry. The example of green plants indicates some of the results that can be expected from molecular nanotechnology:

- Low-cost production of solar collectors
- Low-cost production of large structures (though stronger than wood)
- No production or disposal of toxic chemicals
- Absorption of atmospheric carbon dioxide
- Compatibility with the natural world

Although no technology can, by itself, solve environmental problems, a technology with these characteristics can be a great help. If a high standard of living and reduced environmental impact can be achieved with relatively little sacrifice, then any given amount of political and regulatory pressure should yield greater results in reducing the impact of human activities on the natural world.

Taking the biological analogy as far as the preceding paragraphs have done risks the misunderstanding that molecular nanotechnology will be a form of biotechnology. The differences are large: Molecular nanotechnology will use not ribosomes, but robotic assembly; not veins, but conveyor belts; not muscles, but motors; not genes, but computers; not cells dividing, but small factories making products—including additional factories. What molecular nanotechnology shares with biology is the use of systems of molecular machinery to guide molecular assembly with clean, rapid precision.

Another biological analogy seems appropriate: Aircraft and birds share some basic principles of flight, and birds inspired the development of mechanical flight. It would have been futile, however, to attempt to develop aircraft by applying genetic engineering to birds, or by concentrating exclusively on ornithological research. The Wright brothers studied birds, but they then set off in a fresh direction. Molecular nanotechnology cannot be achieved by tinkering with life, and its products will differ from biological organisms as greatly as a jet aircraft differs from an eagle.

RANGE OF APPLICATIONS

Molecular nanotechnologies will be based on molecular manufacturing, a fundamentally new way to produce materials and devices from simple raw materials. By guiding the assembly of molecules with precision, it will enable the construction of products of unprecedented quality and performance. Because it will work with the fundamental molecular building blocks of matter, it will be able to make an extraordinarily wide range of products.

Computers provide an analogy. In the early decades of this century, many specialized data processing machines were in use: these included the Hollerith punched-card tabulators used in the census, Vannevar Bush's analogue machine that solved differential equations for scientists, and adding machines used in offices to speed accounting chores. Each of these slow, inefficient, specialized machines has now been superseded by fast, efficient, general-purpose computers; even pocket calculators contain computers. By treating data in terms of fundamental building blocks—bits—general purpose computers can perform essentially any desired operation on that data.

Today, manufacturing relies on many specialized machines for processing materials: blast furnaces, lathes, and so forth. Molecular nanotechnology will replace these slow, inefficient, specialized (and dirty) machines with systems that are faster, more efficient, more flexible, and less polluting. As with computers and bits, these systems will gain their flexibility by working with fundamental building blocks. When desktop computers replaced adding machines, they did more than speed addition. Molecular manufacturing will likewise open new possibilities.

The applications of precise fabrication at the molecular level (mechanosynthesis) are as broad as technology itself, because all of technology relies on manufacturing. Molecular-scale components can be used to place the equivalent of a billion modern computers in a desktop machine. Molecular-scale components will make possible new medical and scientific instruments, including DNA readers able to sequence genomes routinely. On a larger scale, production of better materials will make possible lighter, more efficient vehicles, without sacrificing structural strength: this will aid transportation technologies ranging from spacecraft to automobiles. Lighter structures will consume less material and energy. Because the lightest and strongest materials will be made from carbon (in the form of graphite and diamond fibers), carbon dioxide can become a raw material rather than a waste product.

Molecular manufacturing systems can be used to make more molecular manufacturing systems, hence the capital cost of production can be low. An analysis of inputs, outputs, and productivity suggests that the total cost of production can be in the range familiar in agriculture and in the production of industrial chemicals—tens of cents per pound. At this cost, many applications become practical. For example, solar photovoltaic cells fabricated in the form of tough sheets for roofing and paving could provide solar electric power without consuming additional land.

With clean solar power, clean manufacturing processes, and light, efficient products, it will be possible to provide a high material standard of living with decreased impact on the natural world. This can contribute to the goal of sustainable development.

RESEARCH DIRECTIONS AND FUNDING

These developments are not around the corner, but their feasibility can be clearly foreseen, as can the nature of research programs able to implement them. The essential goal is to construct molecular structures with the precision already familiar in chemical synthesis and protein engineering, but on a larger scale. Accordingly, properly focused research in chemical synthesis and protein engineering (within the fields of molecular biology and biochemistry) is important to the implementation of molecular nanotechnology, as is the emerging field of molecular manipulation using proximal probe microscopes such as the scanning tunneling and atomic force microscope.

Each of these areas is a classic small-science field, in which small teams use inexpensive materials and equipment. The prospect of molecular nanotechnology shows that small science can have big rewards.

I have not requested and do not anticipate a need for Federal funds to support my own studies in this area, but the field as a whole could benefit from vigorous support of appropriate computational simulation and laboratory research. Since this work would be performed chiefly by existing researchers with existing equipment, the need is more for a shift in direction than for a growth in spending. Developments along the path to molecular nanotechnology promise to yield early results in scientific instrumentation, making it justifiable as a means of pursuing existing goals in chemistry and in biomedical research.

Progress toward molecular nanotechnology in the U.S. has been retarded chiefly by cultural obstacles. Molecular nanotechnology will require the construction of complex molecular machines, but chemistry and biochemistry are sciences, and focus on the study of nature. To return to the example of aerospace engineering, expecting molecular scientists to build molecular manufacturing systems is somewhat like expecting ornithologists to build aircraft. Building complex systems demands research that first defines goals and then works backward to identify and implement the means, usually dividing the work among many teams. Studying nature, in contrast, can be performed by small research groups, each jealously guarding the independence and purity of its research. The development of molecular nanotechnology can keep much of the character of small science, but it will require the addition of a systems engineering perspective and a willingness on the part of researchers to choose objectives that contribute to known technological goals. Progress will require that researchers build molecular parts that fit together to build systems, but the necessary tradition of design and collaboration—fundamental to engineering progress—is essentially absent in the molecular sciences today.

Furthering molecular nanotechnology might best be achieved by directing federal agencies that perform or fund research in the molecular sciences to support efforts aimed at the construction of molecular machine systems and instruments that can precisely position molecules. The results of this initiative could lead to cost savings in other programs. It has been proposed, for example, that thousands of researchers be employed over many years at great expense in order to read the human genome, yet the molecular machinery found within a dividing cell reads (and copies) the entire genome in a matter of hours. Scientific instruments based on relatively simple molecular machines could read DNA with comparable speed and store the results in a computer memory. The development of such instruments, once the necessary technology base is in place, could hardly consume the efforts of thousands of researchers; it would more likely require only a few cooperating laboratories. The result would enable scientists to read and study many genomes.

Molecular machinery is a technology of basic importance and deserves to be treated accordingly. This would be true even without the longer-term goal of molecular manufacturing.

[A diagram of a planetary gear systems made of 3,557 atoms is inserted at this point in the record.]

RESEARCH IN THE U.S., JAPAN, AND EUROPE

The U.S. has impressive strengths in areas of science and technology relevant to molecular nanotechnology. It was at IBM's Almaden laboratory that Donald Eigler's group spelled "IBM" using 35 xenon atoms. It was at William DeGrado's laboratory at DuPont that scientists first designed and built a new protein molecule, containing hundreds of precisely joined atoms. Nanotechnology has become a buzzword, but is often used to describe incremental improvements in existing semiconductor technologies; although of great value in their own right, these are of surprisingly little relevance to molecular nanotechnology. (Micromachine research, often confused with nanotechnology in the popular press, is even less relevant.)

Progress toward molecular nanotechnology in Japan is harder to judge, owing to distance and language barriers, but the Japanese commitment appears impressive. In my visits to Japan, I have received a strikingly warm welcome. MITI organized a symposium around my first visit, at which—despite my many talks in the U.S.—I for the first time met other researchers who were studying molecular machines not only to understand nature, but to build molecular machine systems. On another visit, I spoke at the only scientific meeting on the construction of molecular machine systems that I have attended but did not myself organize. Japan's NHK television network aired a three-hour series this spring, titled "Nanospace," that included interviews with me and material from my work; nothing comparable has appeared on U.S. television.

While exploring a Japanese-language bookstore that I happened across in Tokyo last spring, I found a table with eight books on micromachines and molecular machines, all displayed face on. Half were paperbacks (including conference proceedings containing a summary of a talk I had given in Tokyo two years before), and half contained one or more graphics illustrating molecular machine designs drawn from my work. One of these was a translation of my first book on molecular nanotechnology, *Engines of Creation*. I can with confidence state that no bookstore in the U.S. contains a similar display, because no such set of books exists in the English language.

MITI's commitment of $185 million is a sign of strong interest. In addition, Japan's Science and Technology Agency, through the Exploratory Research for Advanced Technology program, has sponsored a series of efforts in molecular engineering, including the Aono Atomcraft Project, which aims to build semiconductor devices with atom-by-atom control. I recently read that Texas Instruments has established a laboratory with similar goals; the location they chose is Tsukuba, north of Tokyo.

Researchers at Hitachi's Central Research Laboratory last year spelled "Peace 91 HCRL" by removing individual atoms from a surface. Researchers at the Protein Engineering Research Institute in Osaka (no comparable institute exists in the U.S.) have designed and built the largest protein molecules of which I am aware. Nanotechnology has been a serious goal in Japan for longer than it has in the U.S., and is seen as contributing to technologies in greater harmony with the natural world.

I am less familiar with research in Europe, but key technologies (such as the scanning tunneling microscope) have been developed there. Dr. Hiroyuki Sasabe of the RIKEN Institute in Japan tells me that there are several research consortia in Europe doing work on molecular systems, and that he knows of no similar consortia in the U.S.

POLICY ISSUES

Molecular nanotechnology will raise numerous policy issues. In many areas, years of consideration will be necessary before wise policies can be formulated. This section provides only a brief, preliminary survey of a few issues of particular prominence.

Research in molecular nanotechnology will by its nature pose no special risks so long as it remains unable to make large quantities of product. In its early phases, it will most closely resemble a branch of laboratory chemistry, and its chief product will be information. Later, when large scale applications become possible, major regulatory issues will arise. Further work will be necessary to identify these issues, but because molecular manufacturing can be used to produce high-performance systems of many kinds, these issues will surely include arms control.

Because the U.S. has no clear lead in this technology and because large-scale commercial applications are still distant, international cooperation in research may be desirable. Further, because potential long-term applications include weapon systems, a failure to establish cooperative international efforts could lead to dangerous outcomes. These considerations suggest the desirability of a development program involving international cooperation centering on shared global concerns with health and the environment. One possible vehicle for this might be an expanded version of the existing Human Frontier Science Program.

It seems that no special regulatory issues will arise for some time, but this time should be used to gain an understanding of the issues that will emerge as the technology matures. Cooperative development can provide a basis for eventual international controls, for example, of the use of molecular manufacturing in arms production.

EVALUATING MOLECULAR NANOTECHNOLOGY

The U.S. scientific community has reached no consensus regarding the prospects for molecular nanotechnology; indeed, these ideas have stirred heated controversy. A recent OTA study could identify no published scientific arguments on the other side (vague and unscientific objections have been common), but it would be unwise for a decision maker to advocate a major commitment of resources to molecular nanotechnology without further study and evaluation.

This autumn, the first quantitative, detailed, book-length analysis of molecular manufacturing will be published (*Nanosystems: Molecular Machinery, Manufacturing, and Computation*, Wiley Interscience). This work lays out the fundamental principles of molecular machinery and describes how molecular machines can collect, orient, process, and assemble molecules with high efficiency and reliability. If there is a major error or omission in this analysis of molecular manufacturing, it should be possible for a critic to describe the difficulty in quantitative, scientific terms.

Experience shows, however, that the scientific community does not move swiftly to evaluate interdisciplinary engineering proposals. No single discipline sees it as a responsibility, and most scientists see the work as a distraction from winning their next grant. If these concepts are to be evaluated soon, and well enough to enable decision makers to choose with confidence, deliberate action seems necessary. A natural choice would be to commission a study of molecular manufacturing, setting the objective of evaluating its scientific and technological feasibility by seeking specific, scientific criticisms and responses from appropriate researchers.

A study of this sort could provide a basis for decisions and could stimulate further debate and analysis that would provide a still better basis for decisions. The Office of Technology Assessment may be an appropriate agency to conduct this initial study.

CONCLUSION

Molecular nanotechnology promises a fundamental revolution in the way we make things, and in what we can make. By bringing precise control to the molecular level—resembling the control found in living organisms—it can serve as a basis for manufacturing processes cleaner, more productive, and more efficient than those known today. Like green plants, it can produce inexpensive solar collectors and other useful products while removing carbon dioxide from the atmosphere.

Because it will work with the basic building blocks of matter, its applications are extraordinarily broad: they include improved materials and computers. Early applications will include scientific and medical instruments.

Pure science has prepared the ground for molecular nanotechnology: it is now time to build. Initial goals include the development of better techniques for positioning molecules and for building molecular machines. Research in chemistry, biochemistry, and proximal probe microscopy can all make substantial contributions. Computational simulation has begun to show in detail what can be built and how it will work. Design, simulation, and laboratory research can all benefit from support targeted on genuinely relevant research. Progress will depend largely on the willingness of molecular scientists to solve problems that contribute to engineering objectives.

Research leading toward molecular nanotechnology is accelerating world wide. Focused research is perhaps strongest in Japan. Although large-scale capabilities (and the need for regulation) are still years away, it is not too early to consider the consequences of success and to build the framework of international cooperation that will be necessary in order to manage those consequences.

The preceding paragraphs assume that the analysis supporting the case for molecular manufacturing is essentially correct, but there is as yet no consensus on this. The evaluation of interdisciplinary proposals is slow in the absence of a deliberate effort. It is time to make that deliberate effort, to evaluate the evidence and set research priorities accordingly. If we merely wait and see, we will accomplish more waiting than seeing. Economic competitiveness and the health of the global environment may depend on timely action.

Source: *New Technologies for a Sustainable World*, Hearings before the Subcommittee on Science, Technology, and Space of the U.S. Senate Committee on Commerce, Science, and Transportation, June 26, 1992, pp 102–967.

Position Paper on Nanotechnology in Space Applications

The Molecular Manufacturing Shortcut Group (MMSG) is a Special Interest Chapter of the National Space Society. MMSG is interested in exploring and promoting the applications of nanotechnology to problems of space exploration. The organization has prepared the following position paper on this topic.

POSITION

The National Space Society believes that developing molecular nanotechnology will advance the exploration and settlement of Space. Present manufacturing capability limits the performance, reliability, and affordability of space systems, but the bottom-up approach of molecular nanotechnology has the potential to produce space hardware with tremendous improvement in performance and reliability at substantially lower cost.

BACKGROUND

Molecular nanotechnology "expresses the concept of ultimately being able to arrange atoms in a predetermined fashion by manipulating individual atoms" [Aono]. Its principles were first espoused by Nobel prize winner Richard Feynman in 1959, when he said "the principles of physics, as far as I can see, do not speak against the possibility of maneuvering things atom by atom" [Feynman]. As an engineering discipline, molecular nanotechnology promises revolutionary advances not only in manufactured products, but in the processes used to make them. It is the culmination of many fields:

- Microtechnology strives to build smaller devices;
- Chemistry strives to synthesize more complex molecules;
- Molecular biology strives to manipulate with greater precision the wide range of molecular phenomena that occur in living organisms;
- Materials science strives to make stronger, lighter, and more useful solids; and
- Manufacturing strives to build better products for lower cost.

Each of these fields reaches its ultimate in precise, molecular control, which is the ability to build large structures to complex, atomic specifications by direct positional selection of reaction sites [Drexler1]. These systems should be able to assemble any configuration of atoms, limited only by the laws of nature and human knowledge—hence they are called universal assemblers. Because these assemblers would themselves be made of atoms, and because they would be able to assemble these atoms in arbitrary ways, they should be able to self-replicate, or make copies of themselves. NASA, SSI (Space Studies Institute) and others have recognized the potential impact of applying self-replication to space exploration and development [NASA, SSI, Merkle], but have found that self-replication is difficult for macro-molecular devices, partially because each subcomponent level must deal with errors caused at lower sublevels [Neumann, Toth-Fejel, Long] <1>.

Space exploration and development has benefitted enormously from the advances in these fields, especially microelectronics and materials science, because they reduce payload mass and because they improve reliability. As we converge on the ability to control matter with atomic precision, space development can probably become one of the first and foremost beneficiaries.

Molecular nanotechnology can be confused with the micromachines being produced by microlithographic processes, but the two are very different <2>.

RECENT RESEARCH

The principles of molecular nanotechnology are being demonstrated daily in government and industry laboratories world wide, including the arrangement of 35 xenon atoms to spell out "IBM" [Eigler], the construction of three-dimensional structures from DNA [Seeman], and the engineering of branched, non-biological protein with enzymatic activity [Hahn]. Computer software designed for aiding the development of molecular nanotechnology is also proceeding through the use of tools such as computer-aided design and modeling software [Merkle2].

POTENTIAL BENEFITS AND RISKS

Near Term Benefits

Since the settlement of Space is not a near-term endeavor, it would be a grave mistake to consider only the short term applications of molecular nanotechnology to Space, though there may be a few. In the near term, the chief benefits would most likely be in basic research. For example, improved scanning probes similar to Scanning Tunneling Microscopes (STM) could give researchers a powerful, general technique for characterizing the atomic structure of molecular objects. Such capabilities would be valuable in discovering and designing stronger materials, faster and smaller electronics, and exotic chemicals with unique properties. These incremental improvements would offer the possibility of small improvements in capability across the broad spectrum of space activities, ensuring mission completion, prolonging spacecraft life, and fostering the safety of human crews.

As nanosystems used in research are constructed and commercialized, they will move from gathering basic knowledge in laboratories to collecting data in engineering applications<3>. The first applications would be those in which the relatively high cost and limited capabilities of these first generation devices will still provide significant improvements in overall system capability to justify the costs. Since sensors and actuators could be significantly reduced in size and mass, planetary probes and other space-based applications would probably be one of the first beneficiaries of these nanosystems.

Medium Term Benefits

In the medium term, the nanosystem devices would be directly involved in the manufacturing process. Products might include bulk structures such as spacecraft components made of a diamond-titanium composite, or other "wonder" materials. The theoretical strength-to-density ratio of matter is about 75 times that currently achieved by aerospace aluminum alloys, partially because current manufacturing capability allows macro-molecular defects that weaken the material. The bottom-up approach promises to virtually eliminate these defects, enabling the fabrication of stronger materials that could improve reliability and reduce spacecraft dry weight, resulting in increased payload capacity and higher orbital altitude, ultimately reducing the cost to orbit [DrexlerJBIS].

In the electronics arena, devices might use a few atoms to store a bit of information (as already demonstrated at IBM [Eigler2]). In addition, VLSI (Very Large Scale Integration) would shrink by three magnitudes and extend in three dimensions instead of just two. At this stage, molecular nanotechnology would likely continue to improve capabilities, increase reliability, and lower costs in a wide variety of space projects.

These projected advances would expand the complexity/reliability trade-off envelope for orbital and lunar systems. Tiny, inexpensive inertial guidance systems could assist unmanned exploratory spacecraft, planetary rovers, and interplanetary probes. A dense network of distributed embedded sensors throughout a manned or unmanned spacecraft could continuously monitor (and affect, if they could be operated as actuators) mechanical stresses, temperature gradients, incident radiation, and other parameters to ensure mission safety and optimize system control. In an advanced spacecraft, the outer skin would not only keep out the cold and the vacuum, but it might also function as a multi-sensor camera and antenna. With such extensive monitoring and increasingly efficient control of propulsion systems, life support, and other spacecraft systems, mission success rates would increase at lowered cost. Advanced materials may also enhance on-orbit human activities by providing more effective spacesuits, and may foster more extraterrestrial endeavors by developing more efficient and degradation-resistant solar cells.

As capabilities increase, the molecular techniques used in the actual manufacturing of the spacecraft (i.e. the tools and processes that transform raw materials into advanced sensors and materials) would themselves increase in capability. This advance would make it much easier to build spacecraft systems that could take advantage of in-situ extraterrestrial resources.

Long Term Benefits

Since the settlement of Space is a long term enterprise, these long-term benefits of molecular nanotechnology are the most relevant. And these benefits are considerable. The most important arises from the general ability to build nanosystems, especially the ability to bootstrap production via self-replicating universal assemblers. This capability would probably lower manufacturing costs by many magnitudes, down to the order of $1 per kilogram. It would also make possible to build tapered tethers from geosynchronous orbit to the ground, and to build human-rated SSTO vehicles with a dry mass around sixty kilograms [DrexlerJBIS]. Such capabilities should make possible inexpensive access to space. Mature nanosystems might make possible affordable and robust closed environment life-support systems that could take advantage of in-situ resources, such as asteroidal metals and cometary organics. Such a capability would potentially enable many people to affordably live in space. Tiny computers, sensors and actuators, trivially cheap on a per-unit basis, may allow things like smart walls to automatically repair micrometeorite damage, comfortable and unobtrusive space suits, and terraforming tools. By providing instrumentation that allows the development of medical knowledge at the molecular level, advanced nanosystems might enable in vivo repair of cellular damage. This capability should mitigate the dangers of ionizing cosmic radiation.

Further long term effects of this technology are completely unpredictable, but would undoubtedly be quite significant. Absolute and relative costs will still constrain space activities, however, and some desired activities will remain impossible.

COSTS AND RISKS

Before applications can be developed for the exploration and development of Space, molecular nanotechnology itself must become a practical discipline instead of just a theoretical one. There are three promising paths to the building of universal assemblers: genetic engineering, physical chemistry, and scanning probe microscopy. Uncertainty remains as to which path is easiest and quickest, and hybrid approaches appear quite promising, so efforts should be spread across these three areas. There are likely to be a very large number of expensive blind alleys, so it is important to not invest too much money in any one area.

Japan is aggressively pursuing molecular nanotechnology by investing approximately $200 million over 10 years with industry matching government funding in over twenty companies, while funding for the Atomcraft Project [Aono] is being continued by six Japanese companies as it completes its five year plan. Attempts are underway to start a similar program in Switzerland. If the U.S. fails to engage in activities leading to expanded and well-implemented research with commercially relevant goals, we will probably find ourselves critically behind in the broader economic, military, and specific space-related benefits that may accrue from these technologies. The cost of trailing behind in this technology would be very high.

Many potential threats consist of someone using molecular nanotechnology for aggressive purposes. Thus, efforts must be undertaken to ensure that both global security and U.S. national security are safe against this potential threat. One strategy for ensuring U.S. and global security is to develop molecular nanotechnology in a collaborative, multi-lateral manner. This addresses fears of many nations that they will be caught behind in development, and maintains trust since open collaboration is de facto open and mutual inspection.

There is a possibility that due to some unforeseen law of science, universal assemblers may be impossible to build. In this case, the risk consists of a zero return on investment. But in the ten years since the concepts of molecular nanotechnology have been made public, no one has proposed any scientific reasons for its impossibility. The absence of these reasons might be explained by the fact that numerous objects around us (all carbon-based life-forms) have been formed using the bottom up approach, and by the fact that long-range trends in technology show a continually increase in the precision with which matter can be controlled.

DISCUSSION: SPACE VERSUS
MOLECULAR MANUFACTURING

There is a fear that spending money on molecular nanotechnology will reduce the amount of money spent on Space development, since research funding is sometimes perceived as a zero sum game. One version of this argument asks why the small amount of money available for research should be spent on speculative ventures such as molecular nanotechnology when projects such as DC-X seem to be much closer to success.

- First, decision theory and experience show that achieving large projects of significant technological complexity (e.g. the settlement of Space) require a diversification of effort. It is especially important to have a diversified portfolio of approaches so that unforeseen dead ends can be circumvented without delay. In this case, Space development can benefit significantly by investing a limited amount of effort in low cost, high risk, and high payoff avenues such as molecular nanotechnology.
- Second, the amount of money needed at this stage of molecular nanotechnology development is very small compared to the average NASA Space project.
- Third, much of basic science research, especially the biological and material sciences, concerns itself with the behavior of increasing small groups of atoms. In addition to having nothing in common with Space development, almost all of this research focuses on science, not engineering <4>. Since this research is occurring anyway (regardless of the NASA's and DOD's budget, and hence not part of a zero sum game) it seems that a small effort could steer it in directions most likely to create a space-faring civilization.
- Fourth, the settlement of Space, like the development of molecular nanotechnology, is a long term enterprise, so care must be taken not to sacrifice the long term goals of the latter to achieve short term goals of the former, especially when advanced nanosystems could significantly increase our capabilities in Space.

The opposite version of the "space versus molecular manufacturing" argument ask why money should be spent on expensive Space hardware when exploring and developing Space would be less costly with advanced molecular nanotechnology. While it is true that nanosystems could significantly lower the cost of Space missions, other factors must be considered.

- First and most important, policy makers will undoubtedly make decisions in the near future about molecular nanotechnology, and these decisions should be informed by an acceptance of permanent human presence and expansion into the high frontier of Space. If these decisions are instead made with the assumption that humanity is limited to Earth, the results will most probably be catastrophic.
- Second, it is not known how quickly nanosystems will reach maturity, nor how much effort will be directed toward including them in the design of Space applications. Therefore it seems prudent to continue Space activities and utilize nanotechnologies as they come on line.
- Third, the absence of a significant human direction toward Space may allow social inertia (including cultural attitudes toward frontiers, civil

and criminal law, and levels of technical education) to become a major obstacle in developing nanosystems for Space applications.

- Fourth, Space provides a frontier in which advances in molecular manufacturing will mature with many social impacts. If those social impacts occur in the context of a closed society, it is likely that the Western enlightenment values of humanism, reason, and science will die [Zubrin] <5>.

CONCLUSION

In conclusion, the National Space Society believes that since the settlement of Space is a long range project that will benefit the entire human race, the serious development of the long range field of molecular nanotechnology must be supported. Extraterrestrial activities are a natural application for nanosystems, and synergistic effects between Space and Molecular Nanotechnology can and should be encouraged.

Thanks to the following people for their input and constructive criticism: Max Nelson, Jamie Dinkelacker, Scott Pace, Glenn Reynolds, Bill Higgins, Keith Henson, Craig Presson, Chris Peterson, Jim Bennett, K. Eric Drexler, and the loyal opposition, Allen Sherzer.

REFERENCES

Aono, Masakazu, "Atomcraft," JPRS-JST-92-052-L, 22 June 1992.

Feynman, Richard, "There's Plenty of Room at the Bottom," Engineering and Science, California Institute of Technology, 1960.

K. Eric Drexler, Nanosystems: Molecular Machinery, Manufacturing, and Computation, John Wiley and Sons, 1992.

John von Neumann, Self-Replicating Automata, edited and completed by Arthur Burks, University of Illinois Press, Urbana, 1966.

Toth-Fejel, Tihamer, Self-Test: From Simple Circuits to Self-Replicating Automata, Master's Thesis, University of Notre Dame, 1984.

Long, James and Healy, Timothy, Directors, Advanced Automation for Space Missions, Technical Summary and Report, NASA Ames/University of Santa Clara Summer Study, 1980, (Robert Frietas edited the chapter on space-based applications of self-replication. Of all applications studied, participants believed that this application would have the highest payoff).

Merkle, Ralph, Self Replicating Systems and Molecular Manufacturing, Journal of The British Interplanetary Society, Vol. 45, No. 10, October 1992.

Maryaniak, Greg, Self-Replicating Machines for Space, SSI Update: The High Frontier Newsletter, May/June 1985, Space Studies Institute, Princeton, NJ.

Eigler, D. and E. Schweizer, Positioning single atoms with a scanning tunneling microscope. Nature. 344:524–526, 1990.

Seeman, N., Construction of Three-dimensional Stick Figures from Branched DNA. DNA Cell Bio., 10:475–486, 1991.

Hahn, K., W. Kliss, and J. Steward. Design and Synthesis of a Peptide Having Chymotrypsin-Like Esterase Activity. Science, 248:1544–1547, 1991.

Merkle, Ralph, "Computational Nanotechnology" Nanotechnology Vol. 2, No. 3, pp. 134–141, 1991.

K. Eric Drexler, Molecular Manufacturing for Space Systems: An Overview, Journal of The British Interplanetary Society, Vol. 45, pp. 401–405, 1992.

Eigler, Donald, Christopher Lutz, and William Rudge, Nature, August 15, 1991.

Zubrin, Robert, "The Significance of the Martian Frontier," Ad Astra, September/October 1994.

NOTES

<1>: In top-down technologies, as the manufacturing tool no longer directly affects the workpiece, it must rely on indirect means to add or remove portions of subcomponents. When indirect operations increasingly deal with quantized subcomponents (atoms) as if they were continuous (the top-down assumption), errors grow exponentially.

<2>: First, their components differ in scale by a factor of a thousand. Second, while micromechanical systems are built from the top down (as are all manufactured goods today), nanosystems would be built from the bottom-up, as are chemical feed stocks and biological systems. Finally, self-replication is much more difficult in microtechnology than in molecular nanotechnology, and therefore it lacks the impact such a capability can bring. Because we can manipulate individual atoms with large tools such as scanning probe microscopes, microtechnology is probably not a prerequisite to molecular nanotechnology.

<3>: By exploiting concepts from other technologies, especially biochemistry and microlithography, the cost of scanning probe microscopy will continue decreasing as capabilities simultaneously increase. In addition, the structures of manufactured bulk chemicals, such as buckytubes, will continue increasing in complexity, possibly allowing switching behavior and other non-linear phenomena. Finally, genetic engineering processes will probably continue becoming more flexible and precise, possibly enabling ribosomal construction of quasi-biological structures increasingly different from natural biology.

<4>: Science discovers what is, while engineering creates what has never been.

<5>: The Turner thesis demonstrated that our western progressive humanist civilization depends on frontiers.

Source: NSS Position Paper on Space and Molecular Technology©, <http://www.islandone.org/MMSG/NSSNanoPosition.html>. Reprinted with permission of the Molecular Manufacturing Shortcut Group, a Special Interest Chapter of the National Space Society.

National Nanotechnology Initiative

In his budget for fiscal year 2001, President Bill Clinton proposed the creation of a new National Nanotechnology Initiative, a cross-disciplinary program to promote research and development in nanoscience and nanoengineering. The Initiative is described in a report published in July 2000, "National Nanotechnology Initiative: The Initiative and Its Implementation Plan." The goals and methodologies of the Initiative are laid out in the Executive Summary, as reprinted.

EXECUTIVE SUMMARY

"My budget supports a major new National Nanotechnology Initiative, worth $500 million. . . . the ability to manipulate matter at the atomic and molecular level. Imagine the possibilities: materials with ten times the strength of steel and only a small fraction of the weight—shrinking all the information housed at the Library of Congress into a device the size of a sugar cube—detecting cancerous tumors when they are only a few cells in size. Some of our research goals may take 20 or more years to achieve, but that is precisely why there is an important role for the federal government."

—President William J. Clinton
January 21, 2000
California Institute of Technology

President Clinton's (fiscal year) FY 2001 budget request includes a $225 million (83%) increase in the federal government's investment in nanotechnology research and development. The Administration has made the National Nanotechnology Initiative (NNI) a top science and technology priority. The emerging fields of nanoscience and nanoengineering—the ability to work at the molecular level, atom by atom, to create large structures with fundamentally new molecular organization—are leading to unprecedented understanding and control over the fundamental building blocks of all physical things. The nanoscale is not just another step towards miniaturization. Compared to the physical properties and behavior of isolated molecules or bulk materials, materials with structural features in the ranges of 1 to 100 nanometers—100 to 10,000 times smaller than the diameter of a human hair—exhibit important changes for which traditional models and theories cannot explain. Developments in these emerging fields are likely to change the way almost everything—from vaccines to computers to automobile tires to objects not yet imagined—is designed and made.

The initiative will support long-term nanoscale research and development leading to potential breakthroughs in areas such as materials and manufacturing, nanoelectronics, medicine and healthcare, environment, energy, chemicals, biotechnology, agriculture, information technology, and national security. The effect of nanotechnology on the health, wealth, and lives of people could be at least as significant as the combined influences of microelectronics, medical imaging, computer-aided engineering, and man-made polymers developed in this century.

The initiative, which nearly doubles the nanoscale R&D investment over FY 2000, supports a broad range of scientific disciplines including material sciences, physics, chemistry, and biology, and creates new opportunities for interdisciplinary research. Agencies participating in the NNI include the National Science Foundation (NSF), the Department of Defense (DOD), the Department of Energy (DOE), National Institutes of Health (NIH), National Aeronautics and Space Administration (NASA), and the Department of Commerce's National Institute of Standards and Technology (DOC/NIST). Roughly 70% of the new funding proposed under the NNI will go to university-based research; funds that will help meet the growing demand for workers with nanoscale science and engineering skills. Nanoscience is still in its infancy and, outside of a handful of examples, only rudimentary nanostructures can be created with some control. It will take many years of sustained investment to achieve many of the NNI's research goals, but that is precisely why there is an important role for the Federal government.

Nanotechnology Research and Development Funding by Agency:

	FY 2000 ($M)	NNI ($M)	FY 2001 ($M)	Percent Increase
National Science Foundation	$97	$120	$217	124%
Department of Defense	$70	$40	$110	57%
Department of Energy	$58	$36	$94	66%
NASA	$5	$15	$20	300%
Department of Commerce	$8	$10	$18	125%
National Institutes of Health	$32	$4	$36	13%
Total	$270	$225	$495	83%

The National Nanotechnology Initiative establishes **Grand Challenges**—potential breakthroughs that if one day realized could provide major, broad-based economic benefits to the United States, as well as improve the quality of life for its citizens dramatically. Examples of these breakthroughs include:

- Containing the entire contents of the Library of Congress in a device the size of a sugar cube;

- Making materials and products from the bottom-up, that is, by building them up from atoms and molecules. Bottom-up manufacturing should require less material and create less pollution;
- Developing materials that are 10 times stronger than steel, but a fraction of the weight for making all kinds of land, sea, air and space vehicles lighter and more fuel efficient;
- Improving the computer speed and efficiency of minuscule transistors and memory chips by factors of millions making today's Pentium IIIs seem slow;
- Detecting cancerous tumors that are only a few cells in size using nanoengineered contrast agents;
- Removing the finest contaminants from water and air, promoting a cleaner environment and potable water at an affordable cost; and
- Doubling the energy efficiency of solar cells.

THE NNI INVESTMENT STRATEGY:

The President's Committee of Advisors on Science and Technology (PCAST) strongly endorsed the establishment of the NNI, beginning in Fiscal Year 2001, saying that "now is the time to act." The PCAST noted the NNI as having "an excellent multi-agency framework to ensure U.S. leadership in this emerging field that will be essential for economic and national security leadership in the first half of the next century."

This initiative builds upon previous and current nanotechnology programs. The research strategy is balanced across five kinds of activities. In addition to the Grand Challenges discussed above, these activities include: fundamental research, centers and networks of excellence, research infrastructure, as well as ethical, legal and social implications and workforce programs:

- **Long-term fundamental nanoscience and engineering research** will build a fundamental understanding and lead to discoveries of the phenomena, processes, and tools necessary to control and manipulate matter at the nanoscale. This investment will provide sustained support to individual investigators and small groups doing fundamental research, promote university-industry-federal laboratory partnerships, and foster interagency collaborations.
- **Centers and Networks of Excellence** will encourage research networking and shared academic users' facilities. These nanotechnology research centers will play an important role in development and utilization of specific tools, and in promoting partnerships in the coming years.
- **Research Infrastructure** includes funding for metrology (measurement science), instrumentation, modeling and simulation, and user

facilities. The goal is to develop a flexible and enabling infrastructure so that U.S. industry can rapidly commercialize the new discoveries and innovations.

- **Ethical, Legal, and Societal Implications, and Workforce Education and Training** efforts will promote a new generation of skilled workers with the multidisciplinary perspectives necessary for rapid progress in nanotechnology. Nanotechnology's effect on society— legal, ethical, social, economic, and workforce preparation—will be studied to help identify potential concerns and ways to address them.

Funding by NNI Research Portfolio:

	Fundamental Research	Grand Challenges	Centers and Networks of Excellence	Research Infrastructure	Ethical, Legal, and Social Implications and Workforce	Total
FY 2000	$87 M	$71 M	$47 M	$50 M	$15 M	$270 M
NNI	$90 M	$62 M	$30 M	$30 M	$13 M	$225 M
FY 2001	$177 M	$133 M	$77 M	$80 M	$28 M	$495 M

IMPLEMENTATION PLAN:

Funding of the recommended R&D priorities outlined above will be conducted by the participating agencies as a function of their mission and contingent on available resources. A coherent approach will be developed for funding the critical areas of nanoscience and engineering, establishing a balanced and flexible infrastructure, educating and training the necessary workforce, and promoting partnerships to ensure that these collective research activities provide a sound and balanced national research portfolio. By facilitating coordination and collaboration among agencies, the NNI will maximize the Federal government's investment in nanotechnology and avoid unnecessary duplication of efforts. The vision, strategy, agency participation, and agency partnerships for the five priorities are described in the full report.

Management:

The NNI will be managed within the framework of the National Science and Technology Council's (NSTC) Committee on Technology (CT). The Committee, composed of senior-level representatives from the Federal government's research and development departments and agencies, provides policy leadership and budget guidance for this and other multiagency technology programs.

The CT's Subcommittee on Nanoscale Science, Engineering, and Technology (NSET) will coordinate the Federal government's multiagency nanoscale R&D programs, including the NNI.

The NSET Subcommittee will coordinate planning, budgeting, implementing, and reviewing the NNI to ensure a broad and balanced initiative. The Subcommittee is composed of representatives from agencies with plans for future participation in the NNI and White House officials. The NSET Subcommittee is co-chaired by the White House National Economic Council (NEC) and a representative from an NNI participating agency as designated by the CT. Subcommittee representatives from agencies have operational authority over nanotechnology research and/or nanotechnology infrastructure within their own agency. The NSET Subcommittee succeeds the Interagency Working Group on Nanoscience, Engineering, and Technology (IWGN) as the primary interagency coordination mechanism. Currently, the NSET members are from DOC, DOD, DOE, Department of Transportation (DOT), Environmental Protection Agency (EPA), NASA, NIH, NSF, and White House offices (NEC, Office of Management and Budget (OMB), and Office of Science and Technology Policy (OSTP)).

Under the NNI, each agency will invest in those R&D projects that support its own mission as well as NNI goals. While each agency will consult with the NSET Subcommittee, the agency retains control over how it will allocate resources against its proposed NNI plan based on the availability of funding. Each agency will use its own methods for inviting and evaluating proposals. Each agency will evaluate its own NNI research activities according to its own Government Performance Review Act (GPRA) policies and procedures.

A National Nanotechnology Coordination Office (NNCO) will be established to serve as the secretariat to the NSET Subcommittee, providing day-to-day technical and administrative support. The NNCO will support the NSET Subcommittee in the preparation of multiagency planning, budget, and assessment documents. The NNCO will be the point of contact on Federal nanotechnology activities for government organizations, academia, industry, professional societies, foreign organizations, and others to exchange technical and programmatic information. In addition, the NNCO will develop and make available printed and other material as directed by the NSET Subcommittee, as well as maintain the NNI Web Site.

The NNCO Director will be an NSTC agency representative appointed by the Associate Director for Technology at the White House Office of Science and Technology Policy, in consultation with the Chair of the NSET Subcommittee and Executive Committee of the Committee on Technology. The NNCO Director reports to the Associate Director, but works in close collaboration with the Subcommittee Chair to establish goals and priorities for NNCO support. Agency detailees and contractors will staff the NNCO. The NNCO's annual funding will be derived from cash and in-kind contributions from agencies participating in the NNI, based on a percentage of each agency's total investment in nanoscale R&D. Beginning in FY 2002 and annually thereafter, the NNCO must submit a budget to the NSET Subcommittee for approval. If the NNCO's proposed budget is 15 percent or greater than the previous year, approval by the Executive Board of the Committee on Technology will be required.

The PCAST recommended in its letter to the President that a non-government advisory committee review the NNI annually to assess progress towards its goals. A report would be provided to the Committee on Technology. The Committee would work with the NSET Subcommittee to address issues raised by the outside advisory committee and to implement changes to the NNI strategy.

Initial discussions on such an advisory committee have been held between the members of the Committee on Technology and the National Research Council.

Coordination:

NNI coordination will be achieved through the NSET Subcommittee, direct interactions among program officers within the participating agencies, periodic management meetings and program reviews, and joint science and engineering workshops. The NSET Subcommittee will coordinate joint activities among agencies that create synergies or complement the individual agencies' activities to further NNI goals. Communication and collaborative activities are also facilitated by the NNI website (http://www.nano.gov) as well as by the agencies' sites dedicated to NNI.

Examples of NNI coordination include identification of the most promising research directions, encouraging funding of complementary fields of research across agencies that are critical for the advancement of the nanoscience and engineering field, education and training of the necessary workforce, and establishing a process by which centers and networks of excellence are selected.

The NNI coordination process began in 1999 with the preparation of the IWGN Research Directions. In the spring of 2000, NSET Subcommittee (formerly IWGN) members took part in planning activities at each agency. In addition, a survey is being conducted in all agencies participating in the NNI to identify opportunities for collaboration and areas where duplication can be avoided. Discussions are being held regarding joint exploratory workshops (such as those on molecular electronics, quantum computing, and nano biotechnology) and agreements on specific interagency funding programs. Improved internal coordination in large agencies, concurrently with interagency collaboration, has also been noteworthy in the planning process.

Examples of major collaborative NNI activities planned by the participating agencies are:

Agency	DOC	DOD	DOE	NASA	NIH	NSF
Fundamental research		x	x	x	x	x
Nanostructured materials	x	x	x	x	x	x
Molecular electronics		x		x		x
Spin electronics		x		x		x
Lab-on-a-chip (nanocomponents)	x	x	x	x	x	x
Biosensors, bioinformatics				x	x	x
Bioengineering		x	x		x	x
Quantum computing	x	x	x	x		x
Measurements and standards for tools	x	x	x		x	x
Nanoscale theory, modeling, simulation		x	x	x		x
Environmental monitoring			x	x		x
Nanorobotics			x	x		x
Unmanned missions		x		x		
Nanofabrication user facilities	x		x	x	x	x

The NSET Subcommittee will reach out to the nanoscience and nanoengineering efforts of other nations. These emerging fields create a unique opportunity for the U.S. to partner with other countries in ways that are mutually beneficial. Potential activities include information sharing, cooperative research, and study by young U.S. scholars at foreign centers of excellence. In addition, the NSET Subcommittee will continue its worldwide survey study, DOD's international field offices will assess the nanoscience investment strategies and commercial interests in their geographies of responsibility.

Time line summary:

Below are the key deliverables in the next five years (Fiscal years 2001–2005). Out-year deliverables depend on regular increases in funding for this initiative.

Deliverable	First Achieved
Begin augmented research and development in fundamental research, grand challenges, infrastructure, education and nanotechnology societal impacts in response to open competitive solicitations and regular program reviews	FY2001
Begin work on teams and centers for pursuing mission agency objectives	FY2001
Establish ten new centers and networks with full range of nanoscale measurement and fabrication facilities	FY2002
Develop new standard reference materials for semiconductor nanostructures, lab-on-a-chip-technologies, nanomagnetics, and calibration and quality assurance analysis for nanosystems	FY2003
Develop standardized, reproducible, microfabricated approaches to nanocharacterization, nanomanipulation and nanodevices	FY2004
Develop quantitative measurement methods for nanodevices, nanomanipulation, nanocharacterization and nanomagnetics; Develop 3-D measurement methods for the analysis of physical and chemical at or near atomic spatial resolution.	FY2004
Ensure that 50% of research institutions faculty and students have access to full range of nanoscale research facilities	FY2005
Enable access to nanoscience and engineering education for students in at least 25% of research universities	FY2005
Catalyze creation of several new commercial markets that depend on three-dimensional nanostructures	FY2005
Develop three-dimensional modeling of nanostructures with increased speed/accuracy that allows practical system and architecture design	FY2005

Source: National Science and Technology Council, Subcommittee on Nanoscale Science, Engineering and Technology. *National Nanotechnology Initiative: The Initiative and Its Implementation Plan.* Washington, D.C., July 2000, pages 13–18.

Nanostructure Science and Technology: A Worldwide Study

The following passage is excerpted from a report published in September 1999 by the World Technology (WTEC) Division of the International Technology Research Institute. The report provides data on trends in nanotechnology research from around the world and is based on many studies conducted by WTEC and the National Science and Technology Council's Committee on Technology's Interagency Working Group on NanoScience, Engineering and Technology. As used in this report, the term *nanotechnology* refers to both "top-down" and "bottom-up" technologies

FINDINGS

There are two overarching findings from this WTEC study:

First, it is abundantly clear that we are now able to nanostructure materials for novel performance. That is the essential theme of this field: novel performance through nanostructuring. It represents the beginning of a revolutionary new age in our ability to manipulate materials for the good of humanity. The synthesis and control of materials in nanometer dimensions can access new material properties and device characteristics in unprecedented ways, and work is rapidly expanding worldwide in exploiting the opportunities offered through nanostructuring. Each year sees an ever increasing number of researchers from a wide variety of disciplines enter the field, and each year sees an ever increasing breadth of novel ideas and exciting new opportunities explode on the international nanostructure scene.

Second, there is a very wide range of disciplines contributing to the developments in nanostructure science and technology worldwide. The rapidly increasing level of interdisciplinary activity in nanostructuring is exciting and growing in importance, and the intersections between the various disciplines are where much of the novel activity resides.

The field of nanostructure science and technology has been growing very rapidly in the past few years, since the realization that creating new materials and devices from nanoscale building blocks could access new and improved properties and functionalities. While many aspects of the field existed well before nanostructure science and technology became a definable entity in the past decade, it has only become a coherent field of endeavor through the confluence of three important technological streams:

1. new and improved control of the size and manipulation of nanoscale building blocks
2. new and improved characterization (spatial resolution, chemical sensitivity, etc.) of materials at the nanoscale
3. new and improved understanding of the relationships between nanostructure and properties and how these can be engineered.

As a result of these developments, a wide range of new opportunities for research and applications in the field of nanotechnology now present themselves. Table ES.1 indicates some examples of present and potential applications with significant technological impact that were identified in the course of this study. Considerable resources are being expended around the world for research and development aimed at realizing these and a variety of other promising applications. Government funding alone approached half a billion dollars per year in FY 1997: $128 million in Western Europe; $120 million in Japan; $116 million in the United States; and $70 million altogether in other countries such as China, Canada, Australia, Korea, Taiwan, Singapore, and the countries of the former Soviet Union.

Table ES.2 presents an overall comparison of the current levels of activity among the major regions assessed (Europe, Japan, and the United States) in the various areas of the WTEC study. These broad areas—synthesis and assembly, biological approaches and applications, dispersions and coatings, high surface area materials, nanodevices, and consolidated materials—constitute the field of nanostructure science and technology. These are the areas around which the study was crafted.

In the synthesis and assembly area, the United States appears to be ahead, with Europe following and then Japan. In the area of biological approaches and applications (Chapter 7), the United States and Europe appear to be rather on a par, with Japan following. In nanoscale dispersions and coatings (Chapter 3), the United States and Europe are again at a similar level, with Japan following. For high surface area materials (Chapter 4), the United States is clearly ahead of Europe, followed by Japan. On the other hand, in the nanodevices area (Chapter 5), Japan seems to be leading quite strongly, with Europe and the United States following. And finally, in the area of consolidated materials (Chapter 6), Japan is a clear leader, with the United States and Europe following. These comparisons are, of course, integrals over rather large areas of a huge field and therefore possess all of the inevitable faults of such an integration. At best, they represent only a snapshot of the present, and the picture is admittedly incomplete.

More detailed findings in each of these major areas are included in the individual chapters of this report, along with additional general findings and observations in Chapter 1. Chapter 8 compares the scope and funding levels for the relevant nanostructure science and technology R&D programs around the world. The appendices give details on the site visits and workshops of the panel: B contains the Europe site reports, C contains notes on workshops held in Germany and Sweden, D contains the Japan site reports, and E contains the Taiwan site reports. Appendix A lists the professional experience of panelists and other members of the traveling team.

Table ES.1
Technological Impact: Present and Potential

Technology	Present Impact	Potential Impact
Dispersions and Coatings	Thermal barriers Optical (visible and UV) barriers Imaging enhancement Ink-jet materials Coated abrasive slurries Information-recording layers	Targeted drug delivery/ gene therapy Multifunctional nanocoatings
High Surface Area Materials	Molecular seives Drug delivery Tailored catalysts Absorption/ desorption materials	Molecule-specific sensors Large hydrocarbon or bacterial filters Energy storage Grätzel-type solar cells
Consolidated Materials	Low-loss soft magnetic materials High hardness, tough WC/Co cutting tools Nanocomposite cements	Superplastic forming of ceramics Ultra-high strength, tough structural materials Magnetic refrigerants Nanofilled polymer composites Ductile cements
Nanodevices	GMR read heads	Terabit memory and microprocessing Single molecule DNA sizing and sequencing Biomedical sensors Low noise, low threshold laser Nanotubes for high brightness displays
Additional Biological Aspects	Biocatalysis	Bioelectronics Bioinspired prostheses Single-molecule-sensitive biosensors Designer molecules

Table ES.2
Comparison of Activities in Nanostructure Science and Technology in Europe, Japan, and the United States

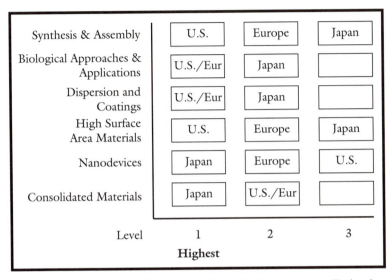

	Level 1 (Highest)	Level 2	Level 3
Synthesis & Assembly	U.S.	Europe	Japan
Biological Approaches & Applications	U.S./Eur	Japan	
Dispersion and Coatings	U.S./Eur	Japan	
High Surface Area Materials	U.S.	Europe	Japan
Nanodevices	Japan	Europe	U.S.
Consolidated Materials	Japan	U.S./Eur	

Source: National Science and Technology Council, Committee on Technology, The Interagency Working Group on NanoScience, Engineering and Technology. *Nanostructure Science and Technology: A Worldwide Study*, Dordrecht, The Netherlands, 1999.
Reproduced by kind permission of the publishers.

Foresight Institute Guidelines

One of the major themes of Eric Drexler's *Engines of Creation* was that the development of molecular nanotechnology (MNT) would bring with it profound social, political, economic, legal, and ethical questions that threatened the basic structure of human society. Indeed, this concern was the driving force behind the creation of the Foresight Institute in 1989. In 1999, a decision was made to prepare a formal statement that would outline the Institute's views as to how research in the area of molecular nanotechnology could go forward with the least risk of imposing unacceptable risk to the general society.

The first draft of the Foresight Guidelines was developed during and after a workshop sponsored by the Foresight Institute and the Institute for Molecular Manufacturing on the weekend of February 19–21, 1999. The Guidelines have gone through many revisions as many people have read, discussed, and commented on the initial draft. The version printed here was released to the general public on June 4, 2000.

The Guidelines are available online on the Foresight Institute's website at http://www.foresight.org/guidelines/current.html. The version printed here and shown on the Institute's website are draft versions. Updated and improved versions will be published at the URL as they become available.

PRINCIPLES

People who work in the MNT field should develop and utilize professional guidelines that are grounded in reliable technology, and knowledge of the environmental, security, ethical, and economic issues relevant to the development of MNT.

MNT includes a wide variety of technologies that have very different risk profiles. Access to the end products of MNT should be distinguished from access to the various forms of the underlying development technology. Access to MNT products should be unrestricted unless this access poses a risk to global security.

Accidental or willful misuse of MNT must be constrained by legal liability and, where appropriate, subject to criminal prosecution.

Governments, companies, and individuals who refuse or fail to follow responsible principles and guidelines for development and dissemination of MNT should, if possible, be placed at a competitive disadvantage with respect to access to MNT intellectual property, technology, and markets.

MNT device designs should incorporate provisions for built-in safety mechanisms, such as: (1) absolute dependence on a single artificial fuel source or artificial "vitamins" that don't exist in any natural environment; (2) making devices that are dependent on broadcast transmissions for replication or in some cases operation; (3) routing control signal paths throughout a device, so that subassemblies do not function independently; (4) programming termination dates into devices, and (5) other innovations in laboratory or device safety technology developed specifically to address the potential dangers of MNT. Further research is needed on MNT risk management, as well as the theory, mechanisms, and experimental designs for built-in safeguard systems.

The global community of nations and non-governmental organizations need to develop effective means of restricting the misuse of MNT. Such means should not restrict the development of peaceful applications of the technology or defensive measures by responsible members of the international community. Further research in this area is encouraged.

MNT research and development should be conducted with due regard to existing principles of ecological and public health.

MNT products should be promoted which incorporate systems for minimizing negative ecological and public health impact.

Any specific regulation adopted by researchers, industry or government should provide specific, clear guidelines. Regulators should have specific and clear mandates, providing efficient and fair methods for identifying different classes of hazards and for carrying out inspection and enforcement. There is great value in seeking the minimum necessary legal environment to ensure the safe and secure development of this technology.

DEVELOPMENT PRINCIPLES

1. Artificial replicators must not be capable of replication in a natural, uncontrolled environment.
2. Evolution within the context of a self-replicating manufacturing system is discouraged.
3. Any replicated information should be error free.
4. MNT device designs should specifically limit proliferation and provide traceability of any replicating systems.
5. Developers should attempt to consider systematically the environmental consequences of the technology, and to limit these consequences to intended effects. This requires significant research on environmental models, risk management, as well as the theory, mechanisms, and experimental designs for built-in safeguard systems.
6. Industry self-regulation should be designed in whenever possible. Economic incentives could be provided through discounts on insurance policies for MNT development organizations that certify Guidelines compliance. Willingness to provide self-regulation should be one condition for access to advanced forms of the technology.
7. Distribution of molecular manufacturing development capability should be restricted, whenever possible, to responsible actors that have agreed to use the Guidelines. No such restriction need apply to end products of the development process that satisfy the Guidelines.

SPECIFIC DESIGN GUIDELINES

1. Any self-replicating device which has sufficient onboard information to describe its own manufacture should encrypt it such that any replication error will randomize its blueprint.
2. Encrypted MNT device instruction sets should be utilized to discourage irresponsible proliferation and piracy.

3. Mutation (autonomous and otherwise) outside of sealed laboratory conditions should be discouraged.
4. Replication systems should generate audit trails.
5. MNT device designs should incorporate provisions for built-in safety mechanisms, such as: (1) absolute dependence on a single artificial fuel source or artificial "vitamins" that don't exist in any natural environment; (2) making devices that are dependent on broadcast transmissions for replication or in some cases operation; (3) routing control signal paths throughout a device, so that subassemblies do not function independently; (4) programming termination dates into devices, and (5) other innovations in laboratory or device safety technology developed specifically to address the potential dangers of MNT.
6. MNT developers should adopt systematic security measures to avoid unplanned distribution of their designs and technical capabilities.

Source: Reprinted by permission of the Foresight Institute. These guidelines available at http://www.foresight.org/guidelines/current.html.

"Technical Boundless Optimism"

Most scientists now seem to agree that researchers will continue to find new and more sophisticated ways of studying matter at the nanoscale level. This research almost certainly will involve the manipulation of atoms and molecules to construct new and interesting structures with a variety of useful properties. But is Eric Drexler's vision of assemblers and replicators really a long-term possibility? Or is *Engines of Creation* nothing more than a work of fantasy, a great excursion in science fiction?

At the dawn of the 21st century, that question still is being debated by many people interested in the field of nanotechnology. Some authorities continue to reject the notion that Drexlerian molecular nanotechnology is worthy of serious consideration. Careful technical rebuttals of Drexler's ideas are rare, perhaps nonexistent. For example, there are few, if any, formal critiques of the arguments presented by Drexler in his book, *Nanosystems*.

One of the most thoughtful presentations of the anti-Drexlerian views is a column written by Dr. David Jones in *Nature* magazine in 1995. The column was a review of Ed Regis' book *Nano!* That article is reprinted here. Jones' article was cited by journalist Gary Stix in his April 1996 *Scientific American* article on molecular nanotechnology. That citation prompted many responses and rebuttals from

supporters of molecular nanotechnology. See, for example, the Foresight Institute webpage on the *Scientific American* controversy
(http://www.foresight.org/SciAmDebate/) and other letters on the
issue, such as http://www.foresight.org/hotnews/Fahy1995. html.

The boundless-optimism school of technical forecasting covers a wide
spectrum. The more realistic end features such marvels as nuclear electricity
too cheap to meter (remember that?) and computers that understand spoken English. At the manic end we find space colonies located at Lagrangian
points, the terraforming of neighbouring planets, and robots in convincing
human form. A promising newcomer has recently appeared somewhere in
the middle of this spectrum. It is called nanotechnology.

Nanotechnology must be carefully distinguished from microtechnology.
The latter, which could operate down to the micrometre scale, is simply the
manufacture and use of extremely small mechanisms—microscopic lathes,
motors and so on. Nobody has got very far with it, and it might or might
not be much use, but the theory seems unexceptionable. As early as 1959
Richard Feynman pointed out that a lathe could be used to make the parts
for a smaller lathe; this could make the parts for one smaller still, and the
process could continue downwards to no very obvious limit. Modern photo-
fabrication can get quite a way in one step, by exploiting the techniques
developed for integrated-circuit manufacture. Micrometre-sized micro-
phones and force-transducers have already been etched out this way.

Nanotechnology takes this idea to its ultimate extreme—the direct assembly of components and structures atom by atom. Its prophets assert that
anything whatever could—and once the trick has been perfected, will—be
made by atomic assembly. The nanotechnological John the Baptist is Eric
Drexler of the Massachusetts Institute of Technology. He has been crying
in the wilderness since at least 1980 and has acquired a group of enthusi-
astic disciples. Regis is one of them, and *Nano!* is his account of the revela-
tion so far.

Nano! is a dramatic work. It is a breathless, exhaustingly complete, des-
perately confusing, utterly uncritical, thought-by-thought, meeting-by-
meeting, paper-by-paper account of Drexler, his life, times and growing
missionary conviction that nanotechnology will lead us to the promised
land. As a story, a personal biography of an imaginative and committed individual, it has its own interest and human appeal. As an account of a clique of
enthusiasts and their vision of the future, it sheds an intriguing light on the
psychology of technical boundless optimism.

The vision, roughly, is this. In the nanotechnological world, everything
will be made by tiny programmable 'assemblers,' about the size of bacteria,

whose exact specification cannot yet be given. Swimming around in a sort of water-bath, they will be capable of assembling atoms into any form whatsoever, including their own. This ability to multiply themselves by replication is important, because vast numbers of them will be needed to construct television sets and rocket motors, atom by atom. They will work rather like the scanning tunnelling microscope—whose stylus can indeed pick up single atoms, move them about and place them in a chosen location. The final products will have no atom out of place. A favoured material of fabrication is diamond. Once the system is up and running, all human material needs will be instantly accessible by simple fabrication.

Even the faithful have qualms about the resulting paradise. How will human beings spend their time, when deprived by universal plenty of their central purpose (seen by Drexler as the competitive display of consumer durables)? What about military and evil applications of nanotechnology? Suppose the assemblers run wild and convert the whole world into intractable grey goo?

But to the unconverted (and it will be obvious by now that I am one of them) the vision is even less convincing. Even if it worked, nanotechnology would not be a universal panacea. How, for example, would it reverse the spreading deserts or halt the population explosion? But the big question, of course, is whether it would work. Regis provides five main arguments in its favour: Feynman has endorsed the feasibility of arbitrarily small machines; scanning tunnelling microscopes have already manipulated single atoms successfully; the normal syntheses of chemistry in effect manipulate single atoms, and do it in trillionfold parallel into the bargain; bacteria function and replicate, so it must be possible; and the uncertainty principle and brownian motion do not seem to worry them.

These arguments deserve closer scrutiny than Regis gives them. How small can a machine, or more precisely a machine tool, be made? It has two essential components: a 'cutter' to act on the workpiece, and a 'frame' to act as a reference, move the cutter around the workpiece and absorb the cutting forces. In a standard engineering machine tool, such as a numerically controlled lathe or milling machine, the cutter is just that; but the analysis applies more generally.

A fundamental requirement of any machine tool is that the frame must be much bigger than the cutter. In a lathe or milling machine, the cutter has to apply its force over such a small area that the local stress deforms or shears the workpiece. The reaction to this force is of the same magnitude, but it must be absorbed by the frame with a deflection that is small compared with the desired resolution of the cutting action. The frame therefore needs a much larger cross-section than the cutter, so that this force represents a very small stress within it. The tool can then impose any pattern of deformation on the workpiece without change or damage to itself. Armed with overwhelming force, it can operate in a completely dictatorial engineering mode.

This requirement for a frame much bigger than the cutter extends to nonmechanical tools as well. A photofabricator, for example, has an optical 'cutter.' To focus as sharply as a wavelength, its lens (by the familiar laws of optics) must be many wavelengths across. To work as an atomic manipulator, a scanning tunnelling microscope might use an electrical potential as its 'cutter.' It would hold and release atoms by applying and removing a strong electric field across them. For this to work, its stylus and stage must be much larger than an atom. A stylus (say) a million atoms long will sustain at most a field one millionth of the one it imposes on the target atom. Now imagine making the microscope smaller and smaller, shrinking it down to a Drexler assembler and even beyond. With every tenfold reduction in size, its internal field rises tenfold. It could perhaps reach the micrometre scale in working order. But by the time it approached the size of the target atom, the field within it would be disrupting its own atoms, and its operation would fail. Once a machine tool is so tiny that its frame cannot be much larger than its cutter, it can no longer function as a tool. Engineering must give way to chemistry, the science of single atoms or groups of atoms acting on each other.

CHEMICAL CONSTRAINTS

And chemistry is a much humbler science than engineering. No chemist can impose an arbitrary design on his raw material, as an engineer can on his. Chemistry is a matter not of dictatorial programming but of cunning and luck and selection. It has to seek out favourable conditions and suitable reagents. It tries to set up molecular encounters that, by their own nature, their random thermodynamic shuffling towards the most accessible local energy minimum and global entropy maximum, will form the desired product. A great many substances and material forms can be made this way, but an infinitely greater number cannot. The successes of chemistry cannot be represented as an argument for nanotechnology.

Indeed, the nanotechnologists do not seem to realize the chemical obstacles in their path. To break a chunk of raw material into its component atoms needs a lot of energy—at least the latent heat of vaporization. And the single atoms when you have them cannot just be picked out and pushed around like so many marbles. When D. M. Eigler and E. K. Schweizer wrote the IBM logo in single atoms in a scanning tunnelling microscope, they used inert xenon atoms in an ultra-high vacuum at liquid helium temperature, dodging the main physical and chemical problems; and even then the atoms stayed put on their substrate only in certain positions and spacings. Single atoms of more structurally useful elements, at or near room temperature, are amazingly mobile and reactive. They will combine instantly with ambient air, water, each other, the fluid supporting the assemblers, or the assemblers themselves. And even if you manage to assemble them into some masterpiece of nanostyling, you had better be sure that it is everywhere an energy-minimum on the atomic scale. Otherwise it won't be stable. Its atoms will recrystallize and diffuse and react towards their own guess at such a minimum. You will have terrible problems with chemistry.

Ah yes, says the nanoenthusiast, but what about biochemistry? What about bacteria, those living exemplars of self-replication and construction on the micro and nano scales? If chemistry is so restricted and troublesome, how do they manage? The answer, I think, is by extraordinary chemical specialization. Life seems to be an exercise in doing the best you can with the minimum of information. The genetic code does not specify the whole structure of an organism atom by atom. It provides a recipe that, if followed without too many blunders, produces a functioning creature capable of handing on much the same recipe. Over evolutionary time, those creatures have prospered whose recipes are the simplest and most robust for what they do.

The same constraint appears in the software of living behaviour. Termites, for example, collectively construct huge nests, in the same sort of way that Drexler imagines his assemblers building a telephone or a dishwasher. Philip Morrison analysed this process instructively in 1979. He showed that the nest results from a mass of termites all blindly following an algorithm with a few simple rules. Nowhere is there any plan or blueprint and no two nests are alike, but the structures always work as nests. It is clear that termite nests, like the termites themselves, have evolved as minimum-information structures. Termites whose nests required a more complicated algorithm failed to survive. Could termites be modified by genetic engineering to assemble, not crude variable nests but (to use Morrison's example) identical working astronomical telescopes? No: the enormous algorithm required would simply overwhelm them.

So it is wrong-headed to argue that, because a creature as small as a bacterium can reproduce itself, therefore a nanotechnological assembler can also replicate itself and, furthermore, will have enough spare capacity to replicate anything else you care to imagine. The chemistry of life centres on some highly specialized organic chemicals: DNA to carry information, a set of amino acids from which to form the proteins that double both as catalysts and as structural material, sugars and fats as energy-stores, and so on.

It is reasonable to guess that this particular biochemistry has survived and stabilized because it operates effectively with the absolute minimum of information. It permits and constitutes the ultimate in elegant programming. To claim that nanotechnological assemblers could work the same chemical trick with chemically unsubtle materials such as aluminium and diamond, and replicate themselves with enough spare capacity to replicate anything else as well, is equivalent to asserting that life itself could be made to work in these materials, and with the added handicap of a vast overhead of arbitrary information.

This, of course, is chemically incredible. Nanotechnology will not be able to avail itself of the sparse, elegant, specialized programming evolved by biochemistry. It will have to work by engineering. Its assemblers will somehow have to pack not only scanning tunnelling microscope technology or its equivalent into the bacterial dimensions, but the terabytes of brute-force coding and processing power needed to specify themselves and their products atom by atom. And I do believe it will fit.

Regis's final argument concerns two imprecisions that lurk in the nano domain, the uncertainty principle and brownian motion. The uncertainty principle, set up and knocked down several times in *Nano!*, is rightly discounted. It doesn't stop atoms having a location and molecules having a structure; it merely makes their edges a bit fuzzy. Only the subatomic level would it start to give serious trouble.

Brownian motion, however, cannot be dismissed so lightly. It raises a fundamental question that the nanoenthusiasts seem to have ignored: that of entropy. Formally this damns their entire project. Above absolute zero, entropic considerations prevent even a single crystal, energy minimum that it is, from ever being perfect. By the same token, the purest product of nanotechnology could never exist at room temperature "with no atom out of place," as Drexler dreams. But let us not be so unforgiving. Crystals are impressively regular structures despite their dislocations, and cells have learnt to live with imperfections in their DNA. A sound energy-minimized nanotechnological design should be able to tolerate a certain proportion of misplaced atoms without disaster. What constraints does thermal dynamics place on its assembly?

DEMON ASSEMBLERS

The very first nanotechnologist was Clerk Maxwell's celebrated Demon, proposed in 1871. This was the little fellow who sat in the connecting pipe between two reservoirs of gas, watching the molecules pass between them. By blocking all molecules travelling one way, but allowing free passage to those going the other way, the Demon could build up a pressure difference across the two reservoirs. By letting fast molecules go one way and slow ones the other, he could build up a temperature difference. Either of these could be used to generate mechanical power, so the Demon could extract energy from a thermal system at one temperature—in contravention of the second law of thermodynamics.

Maxwell's Demon was a puzzle to physicists for many years, but has been well exorcized by now. The first blow was struck by Leo Szilard in 1929, when he pointed out that the Demon needed to obtain information about the molecules. He argued that the entropy generated by this interrogation exactly cancelled the reduction of entropy that could be achieved by exploiting the answer. This elegant analysis initiated our modern identification of information with negative entropy; it has since been taken further and made part of computing theory. It is now clear that Maxwell's Demon cannot possibly work.

Nanotechnological assemblers look suspiciously like Maxwell's Demons. Drexler's argument that they are "programmed, like infecting a bacterium with a virus," evades important questions about their thermodynamics and information flow. How do the assemblers get their information about which atom is where, in order to recognize and seize it? How do they know where they themselves are, so as to navigate from the supply dump to the correct position in which to place it? How will they get their power, for comminution to single atoms, navigation and, above all, for massive internal computing? How will they dispose of the entropy of their operations, and how much will they have to dispose of? The best modern computers still dissipate about 10^{-12} times as much entropy as is theoretically needed by their information flow, so even allowing for improvements it could be quite a lot. Until these questions are properly formulated and answered, nanotechnology need not be taken seriously. It will remain just another exhibit in the freak-show that is the boundless-optimism school of technical forecasting.

Source: David E. H. Jones, "Technical Boundless Optimism," *Nature*, 27 April 1995, 835–37. Reprinted by permission from *Nature* Vol. 374, pp. 835–37. Copyright © 1995 Macmillan Magazines Ltd.

"Why the Future Doesn't Need Us"

As the preceding article suggests, some critics are convinced that some of the claims made for future developments in the area of molecular nanotechnology are technically impossible. Other critics raise a different issue, however. The question, they say, is not whether molecular nanotechnology is possible, but whether such research should even be attempted: Are the potential risks of successful research in this area so great that humans would be better off if it never even took place?

One of the most thoughtful presentations of this position is an article published by Bill Joy in the April 2000 issue of *Wired* magazine.

In that article, Joy, a co-founder of computer giant Sun Microsystems, raises questions about the uninhibited growth of scientific
research in a variety of areas, such as genetics, nanotechnology and
robotics. The following selection describes Joy's concerns about research in molecular nanotechnology in particular.

The many wonders of nanotechnology were first imagined by the Nobel-
laureate physicist Richard Feynman in a speech he gave in 1959, subsequently published under the title "There's Plenty of Room at the Bottom."
The book that made a big impression on me, in the mid-'80s, was Eric
Drexler's *Engines of Creation*, in which he described beautifully how manipulation of matter at the atomic level could create a utopian future of abundance, where just about everything could be made cheaply, and almost any
imaginable disease or physical problem could be solved using nanotechnology and artificial intelligences.

A subsequent book, *Unbounding the Future: The Nanotechnology Revolution*, which Drexler cowrote, imagines some of the changes that might take
place in a world where we had molecular-level "assemblers." Assemblers
could make possible incredibly low-cost solar power, cures for cancer and
the common cold by augmentation of the human immune system, essentially complete cleanup of the environment, incredibly inexpensive pocket
supercomputers—in fact, any product would be manufacturable by assemblers at a cost no greater than that of wood—spaceflight more accessible
than transoceanic travel today, and restoration of extinct species.

I remember feeling good about nanotechnology after reading *Engines of
Creation*. As a technologist, it gave me a sense of calm—that is, nanotechnology showed us that incredible progress was possible, and indeed perhaps
inevitable. If nanotechnology was our future, then I didn't feel pressed to
solve so many problems in the present. I would get to Drexler's utopian future in due time; I might as well enjoy life more in the here and now. It
didn't make sense, given his vision, to stay up all night, all the time.

Drexler's vision also led to a lot of good fun. I would occasionally get to
describe the wonders of nanotechnology to others who had not heard of it.
After teasing them with all the things Drexler described I would give a
homework assignment of my own: "Use nanotechnology to create a vam-
pire; for extra credit create an antidote."

With these wonders came clear dangers, of which I was acutely aware. As
I said at a nanotechnology conference in 1989, "We can't simply do our science and not worry about these ethical issues." But my subsequent conver-
sations with physicists convinced me that nanotechnology might not even
work—or, at least, it wouldn't work anytime soon. Shortly thereafter I
moved to Colorado, to a skunk works I had set up, and the focus of my

work shifted to software for the Internet, specifically on ideas that became Java and Jini.

Then, last summer, Brosl Hasslacher told me that nanoscale molecular electronics was now practical. This was *new* news, at least to me, and I think to many people—and it radically changed my opinion about nanotechnology. It sent me back to *Engines of Creation*. Rereading Drexler's work after more than 10 years, I was dismayed to realize how little I had remembered of its lengthy section called "Dangers and Hopes," including a discussion of how nanotechnologies can become "engines of destruction." Indeed, in my rereading of this cautionary material today, I am struck by how naive some of Drexler's safeguard proposals seem, and how much greater I judge the dangers to be now than even he seemed to then. (Having anticipated and described many technical and political problems with nanotechnology, Drexler started the Foresight Institute in the late 1980s "to help prepare society for anticipated advanced technologies"—most important, nanotechnology.)

The enabling breakthrough to assemblers seems quite likely within the next 20 years. Molecular electronics—the new subfield of nanotechnology where individual molecules are circuit elements—should mature quickly and become enormously lucrative within this decade, causing a large incremental investment in all nanotechnologies.

Unfortunately, as with nuclear technology, it is far easier to create destructive uses for nanotechnology than constructive ones. Nanotechnology has clear military and terrorist uses, and you need not be suicidal to release a massively destructive nanotechnological device—such devices can be built to be selectively destructive, affecting, for example, only a certain geographical area or a group of people who are genetically distinct.

An immediate consequence of the Faustian bargain in obtaining the great power of nanotechnology is that we run a grave risk—the risk that we might destroy the biosphere on which all life depends.

[*At this point, Joy quotes a passage from* Engines of Creation.]

It is most of all the power of destructive self-replication in genetics, nanotechnology, and robotics (GNR) that should give us pause. Self-replication is the modus operandi of genetic engineering, which uses the machinery of the cell to replicate its designs, and the prime danger underlying gray goo in nanotechnology. Stories of run-amok robots like the Borg, replicating or mutating to escape from the ethical constraints imposed on them by their creators, are well established in our science fiction books and movies. It is even possible that self-replication may be more fundamental than we thought, and hence harder—or even impossible—to control. A recent article by Stuart Kauffman in *Nature* titled "Self-Replication: Even Peptides Do It" discusses the discovery that a 32-amino-acid peptide can "autocatalyse its own synthesis." We don't know how widespread this ability is, but Kauffman notes that it may hint at "a route to self-reproducing molecular systems on a basis far wider than Watson-Crick base-pairing."

In truth, we have had in hand for years clear warnings of the dangers inherent in widespread knowledge of GNR technologies—of the possibility of knowledge alone enabling mass destruction. But these warnings haven't been widely publicized; the public discussions have been clearly inadequate. There is no profit in publicizing the dangers.

The nuclear, biological, and chemical (NBC) technologies used in 20th-century weapons of mass destruction were and are largely military, developed in government laboratories. In sharp contrast, the 21st-century GNR technologies have clear commercial uses and are being developed almost exclusively by corporate enterprises. In this age of triumphant commercialism, technology—with science as its handmaiden—is delivering a series of almost magical inventions that are the most phenomenally lucrative ever seen. We are aggressively pursuing the promises of these new technologies within the now-unchallenged system of global capitalism and its manifold financial incentives and competitive pressures.

Source: "Why the Future Doesn't Need Us" © April 2000 by Bill Joy. This article originally appeared in *Wired*, pp. 8–9. Reprinted by permission of the author.

Chapter Seven

Career Information

For most occupations, a person can begin to plan his or her academic career as early as high school or, almost certainly, by the time he or she has reached college. Someone who wants to become a doctor, for example, should take as many life science–related courses as possible in high school and college.

Such is not the case with careers in molecular nanotechnology. Neither high schools nor many colleges offer courses in that subject. Even today, one normally must pursue a graduate degree before encountering courses that carry the term *nanotechnology* in their titles.

The problem of pursuing a career in molecular nanotechnology is difficult for a second reason. All of the most promising and exciting products of the field, such as assemblers, replicators, and nanocomputers, are years or decades into the future. It is possible to find courses that help one to become proficient with molecular modeling, mathematical theory, and other fields that can be used to design and test such devices theoretically. But it will be some time before work on the construction of such objects actually will take place.

Still, there are many fields a person can study that will be essential to the long-range development of a robust molecular nanotechnology, fields that are known as *enabling technologies*. One such enabling technology involves the use of scanning electron and atomic force microscopes. Just as such devices have been used to make some of the most important discoveries in nanotechnology so far, they almost certainly will continue to be crucial in the long-term development of molecular nanotechnology.

Another important field of study is protein engineering. Interest in this subject is growing at a rapid rate because it has the potential for making contributions to such a wide variety of disciplines, including the development of new drugs, the design and manufacture of new "smart" materials, and the growth of molecular nanotechnology itself.

New approaches to chemical synthesis, especially those that use fullerenes, constitute yet another enabling technology. Some of the most exciting breakthroughs in nanotechnology have been made by chemists who have found ways to manipulate individual atoms and molecules, one of the techniques that will be essential in the growth of molecular nanotechnology.

Courses for Nanotechnology

The review of research in molecular nanotechnology cited in Chapter 2 emphasizes one crucial point about this field: its interdisciplinary character. To carry out research in this area, one needs a command of mathematics, physics, chemistry, and biology. High school students interested in a career in molecular nanotechnology should take as many courses in these fields as possible.

Math and science courses are not enough for the prospective researcher, however. As was pointed out in Chapter 2, the social and ethical implications of research in molecular nanotechnology may be profound. While society does not expect research scientists to be authorities in moral and ethical issues, it has a right to expect such workers at least to give some thought to the consequences of their work. Even the most dedicated researcher should have some background in the liberal arts and pursue a well-rounded education in high school and college.

At the college level, more specialized courses relating to molecular nanotechnology are often available. The selection of courses varies widely from institution to institution, but the list available at the University of Washington Center for Nanotechnology illustrates the possibilities. The graduate courses listed here are offered by many different departments within the university, including the departments of physics, chemistry, bioengineering, chemical engineering, electrical engineering, materials science and engineering, biochemistry, molecular biotechnology, and physiology and biophysics. The courses from which one can choose at Washington include:

- Solid state physics
- Spectroscopic molecular identification

- Advanced processing of inorganic materials
- Biomembranes
- Topics in thermodynamics
- Nanotechnology seminar
- Surface analysis
- Technologies for protein analysis
- Polymeric materials
- Laboratory techniques in protein engineering
- Fundamentals of integrated circuit technology
- Physiology seminar
- Introduction to biomechanics
- Biosensors and biomedical sensing
- Frontiers in nanotechnology
- Nanoscale science
- Chemistry and physics of nanomaterials
- Molecular modeling methods
- Molecular self-assembly at surfaces
- Tribology and contact mechanics
- Dyes as molecular probes
- Microelectromechanical systems
- Scanning probe microscopy
- Organic and bioorganic chemistry of nucleic acids in proteins
- Thin film science, engineering, and technology
- Protein machines

Washington is by no means the only institution to have designed programs for students interested in nanotechnology. In Great Britain, School of Industrial and Manufacturing Science of Cranfield University announced a new 1-year program in Microsystems and Nanotechnology beginning October 2001. The syllabus for that program includes, in addition to introductory and background courses, sessions on introduction to microsystems and nanotechnology, functional materials, microsystems technology, microsystems design, advanced microsystems, scanning probe technologies, and molecular nanotechnology.

In many academic institutions in the United States and elsewhere, individual departments or faculty members have developed an interest in molecular nanotechnology and have prepared courses covering some aspect of that field. The July 2000 report on the National Nanotechnology Initiative (see Chapter 11) listed some examples of the kinds of nanotechnology-related courses being offered on U.S. campuses. That list included the following examples:

Advanced Quantum Devices (University of Notre Dame)

Nano-Course (Cornell University)

New Technologies (University of Wisconsin)

Nanostructured Materials (Rensselaer Polytechnic Institute)

Nanoparticles Processes (Yale University)

Nanorobotics (University of Southern California)

Nanotechnology (Virginia Commonwealth University)

Scanning Probes and Nanostructure Characterization (Clemson University)

Nanoscale Physics (Clemson University)

Courses in Nanoscale Science and Engineering

A sample of the courses currently being taught in nanoscale science and engineering is provided. This list comes from Appendix A12 of the 2000 report, *National Nanotechnology Initiative.*

"Advanced Quantum Devices," University of Notre Dame

"Chemistry and Physics of Nanomaterials," University of Washington

"Colloid Chemical Approach to Construction of Nanoparticles and Nanostructured Materials," Clarkson University

"Nano-Course," Cornell University

"Nanoparticles Processes," Yale University

"Nanorobotics," University of Southern California

"Nanoscale Physics," Clemson University

"Nanostructured Materials," Rensselaer Polytechnic Institute

"Nanotechnology," Virginia Commonwealth University

"New Technologies," University of Wisconsin

"Scanning Probes and Nanostructure Characterization," Clemson University

Academic Institutions

As of late 2001, there was only one college or university in the world offering a bachelors degree in nanotechnology: Flinders University in Adelaide, Australia. In addition to basic courses in chemistry, physics, biology, and mathematics, students are required to select among courses such as nanotechnology, professional skills for nanotechnologists, quantum phenomena, introductory microeconomics, molecular biology, solution inorganic chemistry, biodevices, thermodynamics, optics and lasers, chemical bonding and structure, vector calculus, and electrochemistry and kinetics.

Graduate programs in nanotechnology-related topics are more common than those at the undergraduate level. Most of the institutions listed in Chapter 10 as having research programs in nanotechnology also include some type of training component. The Foresight Institute maintains a list of universities that offer courses or degrees in molecular nanotechnology or related subjects (http://www.foresight.org/NanoRev/ FIFAQ2. html). Some of those universities include:

Arizona State University (Center for Solid State Science, Nanostructure Research Group)

Brown University (Nano Micromechanics Laboratory)

University of Cambridge (Cavendish Laboratories and Hitachi Cambridge Laboratories)

Clemson University (The Laboratory for Nanotechnology)

Delft University of Technology (Faculty of Chemical Technology and Materials Science)

Florida Institute of Technology (Division of Engineering Sciences)

Iowa State University (Ames Laboratory Condensed Matter Physics Groups)

Kansas University of Technology (Research Center for Microsystems and Nanotechnology)

Kyushu University (Nano Integration Technology Laboratory)

Michigan State University (The Nanotube Site)

Middle Tennessee State University (Materials Theory and Molecular Design Project)

New York University (Laboratory of Nadrian C. Seeman)

Osaka University (Nanoparticle Group)

Rice University (Center for Nanoscale Science and Technology)

Seoul National University (Center for Science in Nanometer Scale, ISRC)

Stanford University (Stanford Nanofabrication Facility)

Technische Universitat Berlin (Project Nanostructures)

University of Cincinnati (Nanoelectronics Laboratory)

University of Glasgow (Nanoelectronics Research Centre)

University of Hamburg (Nanostructures Physics Group)

University of Lausanne (Nanostructure Research)

University of Leeds (Centre for Nano-Device Modeling)

University of Michigan (Center for Biologic Nanotechnology)

University of Nebraska (Department of Electrical Engineering)

University of Newcastle (Centre for Nanoscale Science and Technology)

University of Southern California (Laboratory for Molecular Robotics)

University of Tokyo (Nanotechnology; Micro-system; Micro-assembly)

University of Toronto (Energenius Centre for Advanced Nanotechnology)

University of Twente (MESA)

University of Washington (Center for Nanotechnology)

Yale University (Department of Engineering)

Jobs in Molecular Nanotechnology

The best source for information about jobs in molecular nanotechnology is often the web page for government, industrial, and academic institutions engaged in nanotechnology research, such as those listed in Chapter 10. Such institutions often have a "Jobs" button on their website that contains information about positions available at the time. Most positions being offered at the time this book was being completed (late 2001) were at academic institutes and included the following:

- Department of Physics, University of Durham (United Kingdom)
- Department of Chemistry, State University of New York at Stony Brook

- Department of Physics, National University of Singapore
- Molecular and Biomolecular Nanotechnology specialty, University of Cambridge (United Kingdom)
- Department of Chemistry, New York University
- Department of Materials Science and Engineering, Cornell University
- Department of Electrical and Computer Engineering, University of Illinois at Urbana–Champaign
- Department of Materials Science and Engineering, University of Michigan
- Nano-Hand Project, Danish Technical University
- Center for Nanoscience and Nanotechnology, Tel Aviv University (Israel)
- Department of Chemical Engineering, University of Akron
- Department of Mechanical Engineering, Iowa State University

Less commonly, job positions were being offered in nonacademic settings, such as the National Aeronautics and Space Administration's Ames Nanotechnology Group. Employment opportunities also were being listed by the first corporation created specifically to promote research in molecular nanotechnology, the Zyvex Corporation. In late 2001, Zyvex was advertising for workers to fill the following positions:

- Physical Chemist, Research Technician, and Surface Scientist in the "Bottom-Up" Division
- Biomolecular Simulation and Molecular Dynamics Simulation Scientists
- Software Architect
- Visualization Tool Developer
- MEMS Processing Engineer
- Research Scientist in Molecular Dynamics
- Laser Spectroscopist

Finally, the Foresight Institute carries occasional mention of job openings in molecular nanotechnology on its website at http://www.foresight.org.

Chapter Eight

Data and Statistics

Private corporations around the world have spent untold billions of dollars on research and development in nanotechnology. Most of those funds have been devoted to traditional "top-down" projects. The amount spent on basic research in molecular nanotechnology is almost certainly small and difficult to determine from corporate reports. Individual researchers and corporate entities are unlikely to distinguish the funds spent on "top-down" versus "bottom-up" research carried out.

Because there are not as yet any commercial products of molecular nanotechnology, no data are available about production, sales, consumption, and other categories normally included in a chapter on data and statistics. Obtaining statistics on nanotechnology in general is difficult because, according to one expert, "there is so much confusion over what the term means. . . . The original notion put forth by K. Eric Drexler (the world's leading authority on nanoscience) is that nanotechnology involves building machines from bottom up with atomic precision. . . . What's happened is that the term has become stylish and has been broadened to apply to many more things." ("Investors Weigh Merits of Nanotech Following Dot-Com Bust." Available at http://www.smalltimes.com/document_display.cfm?document_id= 1214). As a result of this uncertainty and confusion, the only statistics offered here are those dealing with federal financing of nanotechnology in general, particularly that for the National Nanotechnology Initiative proposed by President Clinton in 2000.

National Nanotechnology Initiative Research and Development Funding by Agency, Fiscal Year (FY) 2000–2003 (in Millions of Dollars)

Department/Agency	FY 2000	FY 2001[1]	FY 2002[2]	FY 2003[3]
Department of Commerce	8	33	38	44
Department of Defense	70	125	180	201
Department of Energy	58	88	91	139
Department of Justice	—	1	1	1
Department of Transportation		0	2	2
Environmental Protection Agency	—	5	5	5
National Aeronautics and Space Administration	5	22	22	22
National Institutes of Health	32	40	41	43
National Science Foundation	97	150	199	221
Total	270	464	579	679

Notes: [1] Actual expenditure
[2] Estimate
[3] Proposed

Source: Analytical Perspectives: Budget of the United States Government, Fiscal Year 2003. Washington, D.C.: Government Printing Office, 2002, Table 8-4, page 172.

National Nanotechnology Initiative Funding, Department of Defense, Fiscal Year (FY) 2001–2003 (in Millions of Dollars)[1]

Division	FY 2001[2]	FY 2002[2]	FY 2003[3]
Deputy Under Secretary for Research	36	26	28
Defense Advanced Research Projects Agency	40	97	101
Army	6	20	23
Air Force	10	15	18
Navy	31	22	31
Total	123	180	201

Notes: [1] Data not available for 2000
[2] Actual
[3] Request

National Nanotechnology Initiative Funding, National Science Foundation, Fiscal Year (FY) 2001–2003 (in Millions of Dollars)[1]

Directorate	FY 2001[2]	FY 2002[2]	FY 2003[3]
Biological Sciences	2.33	2.33	2.98
Computer and Information Science and Engineering	2.20	10.20	11.14
Engineering	55.27	86.30	94.35
Geosciences	6.80	6.80	7.53
Mathematics and Physical Science	83.08	93.08	103.92
Education and Human Resources	0	0	0.22
Total, Nanoscale Science and Engineering	149.68	198.71	221.25

Notes. [1] Data not available for 2002
[2] Actual
[3] Request

Source: "National Nanotechnology Investment in the FY 2002 Budget Request by the President."
Available at http://www.nsf.gov/home/crssprgm/nano/2002budget.html.

Chapter Nine

Organizations and Associations

A decade ago, one would have been hard-pressed to find more than a handful of professional organizations whose primary focus was the field of molecular nanotechnology or who even had divisions with such a focus. Today, more and more associations and organizations exist that claim nanotechnology as their primary field of interest or that have established separate divisions for members working in some field of nanotechnology. The following list is by no means complete, although it does give some flavor of the kinds of groups within which scientists, engineers, business people, and others interested in nanotechnology can exchange ideas with other professionals having a similar interest.

American Vacuum Society (AVS)
120 Wall Street, 32nd Floor
New York, NY 10005
Tel: (212) 248-0200
Fax: (212) 248-0245
URL: http://home.vacuum.org/
E-mail: angela@vacuum.org

AVS was organized in 1953 in New York City. A group of 56 people with different backgrounds in physics met to discuss problems and applications of high vacuum technology. The group decided to hold a symposium the following year. At the 1957 symposium, the participants organized themselves as the AVS.

The association has since grown to a membership of 6,000 individuals from around the world. Much of the organization's work is done through its 8 technical divisions, 4 technical groups, and 20 local-area chapters.

In 1991, AVS Section B changed its format to include nanometer-scale science and technology. The subgroup within Section B is called Microelectronic and Nanometer Structure—Processing, Measurement and Phenomena. This subdivision has become increasingly involved not only with nanoscience and nanotechnology in general, but also with molecular nanotechnology in particular. Since 1990, it has been sponsor or co-sponsor of many conferences dealing with scanning tunneling microscope research and related topics. Among these conferences is the International Conference on Nanometer-scale Science and Technology, held annually.

Among the society's activities are short courses at basic and advanced levels, a Science Educators Day designed to inform science teachers about vacuum physics issues, and an Annual Symposium held in the fall of each year. AVS publishes the *Journal of Vacuum Science and Technology*.

One of the most useful resources provided by AVS is an exhaustive listing of all laboratories in which scanning probe microscope work is being conducted. That list is available at http://www.chem.wisc.edu/~nstd/sites3.html.

CMP Cientifica
Apdo. Correos 20
28230 Las Rozas
Madrid, Spain
Tel: +34 91 640 74 40
Fax: +34 91 640 71 86
URL: http://www.cmp-cientifica.com
E-mail: cmpc@cmp-cientifica

CMP Cientifica describes itself as "Europe's only nanotechnology integrator, with activities spanning from basic research, through scientific networks to investment appraisals and due diligence." In the area of research, the organization is presently involved with research projects funded by the European Space Agency and the Spanish Research Ministry.

CMP Cientifica also claims to be a "pioneer is using the Internet to form scientific networks focused on developing applications of nanotechnology." It currently administers three such networks,

PHANTOMS (see below), NanoSpain, and EuroFE, a group of 150 widely scattered researchers working in the area of field emission.

In the area of information, CMP Cientifica issues a weekly e-mail newsletter, "TNT Weekly," to more than a thousand subscribers worldwide. It has also published *Nanotechnology Opportunity Report*, that profiles over 800 companies, research groups, and investors interested in the field of nanoscale science and technology. The organization also organizes conferences for both investors and researchers.

Foresight Institute
Box 61058
Palo Alto, CA 94306
Tel: (650) 917-1122
Fax: (650) 917-1123
URL: http://www.foresight.org
E-mail: inform@foresight.org

The Foresight Institute is a nonprofit educational organization founded in 1989 for the purpose of helping society prepare for anticipated advances in technology. The primary focus of the organization is on the development and potential impact of molecular nanotechnology on society. Its primary officers are individuals who have been involved in molecular nanotechnology since its earliest development, including K. Eric Drexler (Chairman), Chris Peterson (Executive Director), Stewart Brand, Arthur Kantrowitz, Ralph Merkle, and Marvin Minsky (all members of the Board of Advisors).

Three levels of membership are available in the institute. A free electronic membership allows a person to receive a quarterly e-mail update about news from the institute. Regular memberships bring a printed copy of *Foresight Update* in addition to various background and briefing papers and quarterly e-mail announcements. Senior Associate Memberships are available to those who make annual contributions of at least $250 for 5 years. Senior Associates receive special publications, online interactions, and invitations to special Senior Associate meetings.

The institute publishes a quarterly newsletter, *Foresight Update*, designed to inform professionals in the field and the general public about technical and nontechnical developments in nanotechnology. A quarterly electronic newsletter is also sent to members. Since 1989, the institute also has sponsored annual conferences on molecular nanotechnology. The proceedings of two of those conferences have been produced in book form.

Two ancillary organizations have been spun off from the institute, the Institute for Molecular Manufacturing and the Center for Computational Issues in Technology. The former organization is a nonprofit research organization founded to promote research in molecular nanotechnology, and the latter was created as a nonprofit California corporation whose purpose it is to study public policy issues arising as a result of the development of new technologies, such as molecular nanotechnology. Separate entries are provided for each of these organizations.

Institute for Molecular Manufacturing (IMM)
555 Bryant Street, Suite 253
Palo Alto, CA 94301
Tel: (415) 917-1120
Fax: (415) 917-1123
URL: http://www.imm.org
E-mail: admin@imm.org

IMM was formed as a nonprofit foundation in 1991 for the purpose of supporting research in the area of molecular nanotechnology. Its parent group is the Foresight Institute. IMM co-sponsors many activities with the Foresight Institute, including the quarterly newsletter *Foresight Update* and annual conferences on molecular nanotechnology. IMM also publishes *IMM Reports* that carry news about recent developments in research. About half of the issues of *IMM Reports* have been on "Steps Toward Nanotechnology," written by Jeffrey Soreff, and the remaining issues have dealt with topics such as nanomedicine, the private air car, and building molecular machine systems.

The three major objectives of IMM are to "(1) fund seed grants to scientists for research and development; (2) act as an interdisciplinary clearing-house for U.S. and international developments in technologies and processes leading to molecular manufacturing; and (3) disseminate educational materials and sponsor seminars on molecular manufacturing."

Institute of Nanotechnology
9 The Alpha Centre
Stirling University Innovation Park
Stirling FK9 4NJ
United Kingdom
URL: http://www.nano.org.uk/
E-mail: o.saxl@nano.org.uk

The Institute of Nanotechnology functions as an "umbrella group" in the United Kingdom, providing information on a wide range of topics related to nanotechnology. Their website provides news on investing in nanotechnology, announcements of conferences, educational opportunities, biographical sketches of important figures in the field, and information on funding programs.

International Union for Vacuum Science, Techniques, and Applications Union (IUVSTA)

c/o Dr. William Westwood, Secretary General
7 Mohawk Crescent
Nepean, Ontario, K2H 7H7
Canada
Tel: (613) 829-5790
Fax: (613) 829-3061
URL: http://www.vacuum.org/iuvsta/
E-mail: westwood@istar.ca

IUVSTA was formed on June 13, 1958, in Namur, Belgium. It was created to foster international cooperation on research in vacuum science and related interdisciplinary topics. The organization now includes groups from 12 nations with a total membership of about 15,000 physicists, chemists, materials scientists, engineers, and technologists. Typical of the activities sponsored by the organization are the triennial International Vacuum Congress, an annual series of European Conferences on Surface Science, a regular series of European Vacuum Conferences, and a series of workshops on vacuum science and technology and related topics.

Los Angeles Regional Technology Alliance (LARTA)

746 West Adams Boulevard
Los Angeles, CA 90089-7727
Tel: (213) 743-4150
Fax: (213) 747-7307
URL: http://www.larta.org
E-mail: mail@larta.org

LARTA was organized in 1993 through legislation enacted by the State of California as a way of helping to deal with downsizing in the aerospace industry. The organization is a think tank for technology industries involving corporations, investers, government, and researchers. Among the functions of LARTA are market research conducted for both the general public and private clients; a California seed grant fund to support cutting-edge technologies; and "Entrepreneur's toolbox" on the association's website to provide information on technology for businesses; a weekly newsmagazine, La Vox; conferences and seminars through Larta University; and free access to research reports, entrepreneurial advice, and new ideas on the association's website.

Microscopy Society of America

Bostrom Corporation
230 East Ohio, Suite 400
Chicago, IL 60611
Tel: (312) 644-1527/(800) 538-3672
Fax: (312) 644-8557
URL: http://www.msa.microscopy.com
E-mail: BusinessOffice@MSA.Microscopy.Com

The Microscopy Society of America is a nonprofit organization dedicated to the advancement of knowledge in all areas of microscopy, including imaging, analysis, and diffraction techniques in the areas of life sciences, materials, medicine, and the physical sciences. The society is a sponsor with three other societies of the journal *Microscopy and Microanalysis.*

Molecular Manufacturing Enterprises Incorporated (MMEI)

9653 Wellington Lane
St. Paul, MN 55125
Tel: (612) 288-0093
URL: http://www.mmei.org
E-mail: svetter@mmei.com

MMEI is a seed capital firm, that is, an organization created to provide funding for research in the area of molecular nanotechnology. In addition, the company provides advice, contacts, and other support services to researchers and business people involved in research and development in the area of molecular nanotechnology. The organization solicits proposals from for-profit and nonprofit organizations, individuals, groups, and academic institutions. It also may assist individuals and groups with finding funding resources greater than those provided by MMEI itself.

Molecular Manufacturing Shortcut Group (MMSE)

8381 Castilian Drive
Huntington Beach, CA 92646
Tel: (734) 662-4741
URL: http://www.islandone.org/MMSG/
E-mail: ttf@re.net

MMSG is a chapter of the National Space Society. Its mission is to promote the development of nanotechnology as a means to facilitate the settlement of space. Its long-range goal is to inform government, industry, academia, and the space activist community what molecular nanotechnology is and how it can be used to facilitate the development of space. MMSG members receive a quarterly newsletter and have access to presentation materials on molecular manufacturing.

Nanobusiness Alliance
244 Madison Avenue, Suite 485
New York, NY 10016
Tel: (845) 247-8920
URL: http://www.nanobusiness.org
E-mail: info@nanobusiness.org

The Nanobusiness Alliance describes itself as "the first industry association founded to advance the emerging business of nanotechnology and Microsystems." The Alliance was founded in 2001 by leaders from government, industry, and academia. Some of the activities planned for the group include the production of white papers, forecasts, surveys, industrial directories, position papers, analyses of legislation, public awareness campaigns, job banks, mentoring programs, message boards, and capital access initiatives. The Alliance also plans to provide expert testimony to legislative and regulatory bodies and aid in the development of regional nanobusiness centers.

National Science and Technology Council (NSTC)
National Nanotechnology Initiative
Dr. M. C. Roco, Chair
National Science Foundation
4201 Wilson Boulevard, Suite 525
Arlington, VA 22230
Tel: (703) 292-8371
Fax: (703) 292-9054
E-mail: http://www.nsf.gov/nano
URL: http://www.nano.gov/nset.htm

NSTC was established by Executive Order on November 23, 1993. The purpose of the agency is to advise the President on science, space, and technology policies. NSTC's Committee on Technology provides policy leadership and budget guidance for the nation's new National Nanotechnology Initiative. The Committe on Technology's subcommittee on Nanoscale Science, Engineering and Technology is responsible for coordinating the federal government's nanoscale research and development program funded through the NNI. For more detailed information on the organization and operation of these entities, see the full report on the NNI in Chapter 9.

PHANTOMS
c/o CMP Cientifica
Apartado de Correos 20
28230 Las Rozas
Madrid, Spain
Fax: +34 91 6407186
URL: http://www.phantomsnet.com
E-mail: antonio@cmp.cientifica.com

PHANTOMS is an interdisciplinary network of 125 researchers from 17 European countries, the United States, Canada, Japan, and India. It is funded by the European Commission through its IST (Information Sciences Technology) program. The goal of PHANTOMS is to keep Europe on a competitive basis with the United States and Japan in the development of nanotechnology advances. Research areas of interest to the association include modeling, lithography, nanoscale optics, molecular and bioelectronics, nanoprobes, nanoimprinting, and self-assembly.

Society for the Advancement of Material and Process Engineering (SAMPE)
International Business Office
1161 Parkview Drive
PO Box 2459
Covina, CA 91722-8459
Tel: (626) 331-0616
Fax: (626) 332-8929
URL: http://www.sampe.org/
E-mail: sdsampe@znet.com

SAMPE evolved out of the Society of Aircraft Material and Process Engineers, founded in 1944 by nine Southern California aerospace engineers. By 1959, there were 15 chapters of the organization in various parts of the United States. In 1973, the organization broadened its focus and was renamed The Society for the Advancement of Material and Process Engineering.

SAMPE now consists of about 40 professional chapters, 10 of which are located outside the United States. There are 47 student chapter affiliates. Special areas of interest in the organization include aircraft/aerospace/military, composite components, transportation, construction, medical, and appliance/business equipment.

SAMPE's major activities are international and regional symposia and conferences. These meetings deal with topics such as composites, metals, ceramics, fabrication processes, polymers, reinforcements, adhesives, testing and analysis, and environmental issues.

SPIE—The International Society for Optical Engineering
PO Box 10
Bellingham, WA 98225
Tel: (360) 676-3290
Fax: (360) 647-1445
URL: http://www.spie.org
E-mail: spie@spie.org

SPIE is a nonprofit professional society whose goal is to advance research, engineering, and applications in optics, imaging, electronics, and other optical phenomena. As such, the society is concerned with "top-down" research aimed at producing smaller computers and other nanoscale devices.

SPIE produces three major print journals, *Optical Engineering*, *Journal of Biomedical Optics*, and *Journal of Electronic Imaging*. It also sponsors an Internet page, *SPIE Journals Online*. In addition to these technical publications, the society publishes a monthly newspaper, *OE Reports*, that covers not only new and evolving technologies, but also business and economic trends in the industry.

Most of the work in SPIE is done through more than 20 technical groups consisting of professionals and organizations worldwide. Some examples of those groups are Biomedical Optics, Electronic Imaging, Fiber Optics, Optoelectronics, Penetrating Radiation, and Photolithography.

Chapter Ten

Research Organizations

Research in a variety of fields related to nanotechnology now is being conducted at government, industrial, and academic laboratories. In many cases, there is little or no distinction between "top-down" and "bottom-up" types of research. In almost all cases, the more traditional approach to nanoscale research is predominant or the only type of research being conducted at that scale. The institutions listed in this chapter were selected because of their commitment, to at least some extent, to "bottom-up" research in nanotechnology. The distinction among governmental, industrial, and academic laboratories is arbitrary. In many cases, research is supported by two or three of these entities, and it is not clear to which category a research group should be assigned.

The list of research organizations in this chapter should not be considered as complete. Nanoscale science and technology is a rapidly growing field, with new research groups appearing on the scene almost monthly in almost every developed nation in the world. The institutions listed here have been selected because they provide some sense of the kinds of research being conducted in the field and also because additional information about their work can be accessed rather easily on the Internet or by some other means.

Government (Including National Laboratories) and Other Nonprofit Organizations

Air Force Office of Scientific Research
801 North Randolph Street, Room 732
Arlington, VA 22203-1977
URL: http://ecs.rams.com/afosr/
E-mail: afosr-m@rams.com

The Air Force Office of Scientific Research has supported a modest amount of research in the area of molecular nanotechnology with relevance to such topics as solid mechanics and structures, structural materials, and electronics. For example, it provided financial support for research on specially designed nanostructured polymers, large molecules designed and built to perform specific functions, similar to those carried out by proteins in living cells.

Additional information about Air Force research activities is available from TECH CONNECT, at Wright-Patterson Air Force Base, AFRL/XPTT, Building 15, 1864 4th Street, Suite 1, Wright-Patterson AFB, OH 45433-7131, 937-656-9030 or 800-203-6451 or on the web at http://www.afrl.af.mil.

Argonne National Laboratory (ANL)
9700 S. Cass Avenue
Argonne, IL 60439
Tel: (630) 252-2000
URL: http://www.anl.gov/

At the end of 1999, ANL was still defining its precise role in the development of nanotechnology research. ANL is a member of an interagency work group with responsibilities for a national initiative called "Nanotechnology for the Twenty-First Century: Leading to a New Industrial Revolution." Some of the themes to be addressed in this initiative include nanoscale instrumentation, controlled synthesis and processing at the nanoscale, manipulation and coupling at the nanoscale, nanoscale precursors and assembly for macrostructures and devices, understanding and mimicking biological functions, and hybrid subsystems.

Center for Applied Nanotechnology
Industrial Technology Research Institute
195 Chung Hsing Road, Sec. 4 Chu Tung
Hsin Chu, Taiwan 310
Republic of China

Tel: +886-3-582-0100
Fax: +886-3-582-0045
URL: http://www.itri.org/
E-mail: lclee@itri.org.tw

ITRI opened in January 2002 with a pledge of NT$19.2 billion (U.S.$548.7 million) to make Taiwan a top-level player in the world of nanotechnology research. The institute will focus on four major areas: raw materials, electronics, machinery and biomedicine.

Center for Fullerenes and Nanostructures (CFN)
Institute for High-Performance Computing and Data Bases
Ministry of Science and Technology
PO Box 71
194291 Sankt-Petersburg
Tel: (812) 251-00-38
Fax: (812) 251-83-14
URL: http://www.csa.ru
E-mail: kozyrev@fn.csa.ru

CFN was organized to conduct basic and applied research on fullerenes and nanostructures. It uses computer simulations and theoretical calculations on chemical and physical structures and reactions to design new nanomaterials and nanomechanisms.

CFN is divided into three laboratories: Molecular Modeling of Biosystems Laboratory, Laboratory of Nanoclusters and Atomic Structures, and Laboratory of Informational Support for Fundamental Research. Some examples of the type of research currently being conducted at CFN include the development of new tools and methods for molecular modeling and representation of molecules and supermolecules; research on medical applications of fullerenes and nanostructures; quantum chemical calculations of the properties of nanoclusters; and research on diamonoid materials.

Center for NanoSpace Technologies, Inc. (CNT)
PO Box 70
League City, TX 77574-0070
Tel: (281) 334-9610
URL: http://www.nanospace.org/
E-mail: sauty@nanospace.org

CNT is a nonprofit research foundation established for the purpose of designing and carrying out research at the forefront of technological advance. The results of this research are applied to the aerospace, biomedical, educational, petrochemical, and shipping/transportation industries.

Among CNT's principal objectives are the support and encouragement of interdisciplinary research and development to advance the field of microscience and nanoscience; to encourage the interchange of ideas about microscience and nanoscience among engineers, scientists, managers, and policy makers; and to organize and promote conferences and other meetings for the purpose of presenting new information about the microsciences and nanosciences.

Center for Nanotechnology
NASA Ames Research Center
Mailstop 229-3
Moffett Field, CA 94035
Tel: (650) 604-2616
Fax: (650) 604-5244
URL: http://www.ipt.arc.nasa.gov/
E-mail: meyva@orbit.arc.nasa.gov

The NASA Center for Nanotechnology was formerly known as the Integrated Product Team on Devices and Nanotechnology. Its research is classified into one of three major areas: (1) nanoelectronics and computing, (2) sensors, and (3) structural materials. Research in the area of nanoelectronics and computing focuses on the development of molecular electronics and photonics devices, computing architecture, and assembly mechanisms. Sensors are being developed for the detection of life forms and the monitoring of crew health and safety and condition of vehicle health. Research in structural materials is aimed at the development of new types of composites and multifunctional and self-healing materials.

Some potential applications of molecular nanotechnology in space programs might include very small sensors, power sources, and communication, navigation, and propulsion systems for use in small mass and volume spacecraft; "intelligent" spacecraft that can operate essentially on their own; the design of entirely new kinds of space structures; and planetary exploration systems that consist of micro-rovers, microspacecraft, and networks of ultrasmall probes on planetary surfaces.

Much of the research being conducted at the Center for Nanotechnology now makes use of various types and forms of carbon nanotubes that can be used for nanolithography, data processing and storage, structural components, electronic computing, and fuel cells.

Community Research and Development Information Service (CORDIS)

B.P. 2373
L-1023 Luxembourg
Tel: (+352) 44-10-12-2240
Fax: (+352) 44-10-12-2248
URL: http://www.cordis.lu/
E-mail: helpdesk@cordis.lu

CORDIS is the primary source of information concerning research and development programs funded by the European Community. The European Union operates many multiyear programs in research, technological development, and demonstration (RTD) activities. The Fifth RTD Framework Programme covered the period from 1998 through 2002. It consisted of two distinct parts. One part covered RTD activities in general, and the second part focused on research and training in the nuclear sector.

CORDIS is the central clearinghouse for information about RTD activities. It maintains a Help Desk, which can be accessed through its website, and many print publications, such as *CORDIS Focus*, a printed abstract of the news database, and *Euro-Abstracts*, the printed version of the publications database.

A search of the CORDIS database found many references to research in nanotechnology. Most of this research was of the "top-down" type, although examples of "bottom-up" research also were available.

A division within the CORDIS database of particular interest to readers of this book is the proactive initiative called "Nanotechnology Information Devices." This initiative has been soliciting research proposals for "future information processing and storage systems that operate at the atomic or molecular scale in order to achieve superior functionality or performance." One of the specific objectives of the program is the support of research on "tools and techniques for the fabrication of structures with critical dimensions below 10 nm," including topics such as self-organization, self-assembly, directed nanoassembly, and associated microtools or nanotools.

Competence Centre for the Application of Nanostructures in Optoelectronics (NanOp)

Institut für Festköroperphysik
TU-Berlin
Sekr. PN 5-5
Hardenbergstrasse 36

10623 Berlin
Germany
Tel: +49-30-314-79-605
Fax: +49-30-314-75-138
URL: http://www.nanop.de/
E-mail: nanop@sol.physik.tu-berlin.de

NanOp is the national network in Germany of large, medium-size, and small companies; national institutes for applied and fundamental research; and universities involved in research in the areas of nanostructures, nanoanalytical techniques, and optoelectronics. The organization has a single goal, which it describes as "to speed up research in and development of nanotechnologies for radically new proprietary NanOp products, components, and systems based on these components."

The types of products on which NanOp research groups are working include consumer products, such as laser TV, illumination, and displays; computers and communications, such as data storage and optical interconnects; and measurement and monitoring techniques, such as gas sensing and contamination control.

Defense Advanced Research Projects Agency (DARPA)

3701 North Fairfax Drive
Arlington, VA 22203-1714
Tel: (703) 674-5469
URL: http://www.darpa.mil/

DARPA is a separately organized agency within the U.S. Department of Defense. DARPA engages in advanced basic and applied research and development essential to the Department of Defense and its variety of military missions. The Agency arranges, manages, and directs the performance of work carried out by a variety of military and civilian organizations, much of it at the leading edge of research in a variety of scientific fields, including materials science.

For many years, DARPA has supported research in the area of microelectronics and molecular nanotechnology. Examples of current programs administered by the Agency include Molecular Electronics (Moletronics), Advanced Microelectronics (AME), and Molecular-level, Large-area Printing (ML). The purpose of Moletronics is to show ways in which molecules or nanoparticles can be used in electronic devices in such a way as to interconnect with macroscopic equipment. The goal of AME is to explore the possibility of producing electronic devices with dimensions as small as 25 nm that can be used in chips containing up to 10^{12} transistors each. MLP focuses on the development of new techniques for the fabrication and patterning of ultrasmall circuitry on flat and curved substrates.

Exploratory Research for Advanced Technology (ERATO)
Kawaguchi Center Building
1-8, Honcho 4-chome
Kawaguchi City
Saitama Prefecture
332-0012 Japan
Tel: +81 048 226-5621
Fax: +81 048 226-5653
URL: http://www.jst.go.jp/EN/
E-mail: www-admin@tokyo.jst.go.jp

ERATO was established in 1981 by the Japanese government for the purpose of fostering the creation of advanced science and technology while stimulating future interdisciplinary research and searching for better systems through which to carry out basic research. The program sponsors research in many different areas, with each project being funded for a 5-year period at a level of about 1.7 billion yen ($16 million U.S. dollars).

The project that may have had the greatest relevance to molecular nanotechnology was the AONO Atomcraft project, which functioned from 1989 through 1994. Project director was Dr. Masakazu Aono, Chief Scientist at RIKEN, the Institute of Physical and Chemical Research. The term *atomcraft* was coined to describe the project's focus on the manipulation of individual atoms and molecules with scanning probe microscopes. The project's accomplishments are summarized on its web page at http://www.jst.go.jp/erato/project/agsh_P/agsh_P.html.

Some current projects also focus on topics in molecular nanotechnology. For example, research in the NAMBA Protonic NanoMachine project focuses on three major topics: (1) nanoassembly, (2) nanoswitching, and (3) nanomechanics. Information about the project is available on the group's web page at http://www.jst.go.jp/erato/project/npnm_P/npnm_P.html.

Institute for New Materials (Institut für Neue Materialen) (INM)
Im Stadtwald-Gebäude 43
D-66123 Saarbrücken
Germany
Tel: +49-681-9300-312/313/395

Fax: +49-681-9300-223
URL: http://www.inm-gmbh.de/
E-mail: schmidt@inm-gmbh.de

INM was established in 1988 as an entity within the University of the Saarland. Its purpose is to develop new materials that will be needed on a large scale by industry in the future. The focus of research at INM since 1990 has been on the integration of organic synthesis chemistry with chemical nanotechnology. This research has led to the development of new materials in the areas of metals, non-metallic inorganic materials (e.g., glasses and ceramics), organic polymers, and chemical composite materials.

Institute for Solid State and Materials Research Dresden
(Institut für Festkörper und Wekstofforschung) (IFW Dresden)
PF 27 00 16
D-01171 Dresden
Germany
Tel: +49 (0)351 4659 0
Fax: +49 (0)351 4659 540
URL: http://www.ifw-dresden.de
E-mail: postmaster@ifw-dresden.de

IFW Dresden was founded in 1992 and is funded by the Free State of Saxony and the Federal Republic of Germany. It is affiliated with the Wissenschaftsgemeinschaft Gottfried Wilhelm Leibniz and is a member of the Materials Research Network Dresden. Research at IFW Dresden is divided into eight research areas, some of which are directly related to either "top-down" or "bottom-up" nanotechnology. These areas include Thin Film Systems: Mechanical Properties; Thin Film Systems: Electronic Properties; and Conducting Polymers, Fullerenes, and Nanotubes. Two examples of more recent research at IFW Dresden include studies on carbon nanotubes and the formation of nanostructures in C60 films.

Los Alamos National Laboratory (LANL)
Los Alamos, NM 87545
Tel: (505) 667-4869
Fax: (505) 665-4063
URL: http://bifrost.lanl.gov/

Researchers at LANL are attacking problems in nanotechnology at three different levels. First, members of the Condensed Matter and Statistical Physics Theoretical Division are studying some fundamental theoretical problems involved in the development of nanolevel structures.

Second, other researchers are developing nanoscale materials using the "top-down" approach. For example, a team of LANL researchers have been working in a collaborative program with Russian scientists to develop nanocrystalline materials that can be used for orthopedic implants.

Third, other workers at the laboratory are exploring techniques for use in "bottom-up" construction of nanoscale materials. For example, researchers in early 1999 announced that they had developed an ultrafast scanning tunneling microscope with which events taking place in the picosecond (trillionths of a second) range at the atomic level could be detected.

Nano Science and Technology Center

Chinese Academy of Science
(Various locations)
URL: http://www.casnano.net.cn/

The Nano Science and Technology Center is composed of a variety of research centers in China where research on nanoscale science and technology are taking place. Some examples of the dozens of projects underway under auspices of the Center are "Preparation, Physical Properties and Application of Carbon Nanotubes and Their Oriented Arrays," "Carbon Nanotubes and Other Nanomaterials," "Study on Several Key Problems of Nanometer Science and Technology," "Study on Preparation of Nanoscale Information Functional Materials," and "Nanoscale Ceramics and Nanocomposites."

NAS Nanotechnology Team

NASA Ames Research Center
Mail Stop 258-5
Moffett Field, CA 94035-1000
Tel: (650) 604-5026/(650) 604-9000
URL: http://www.nas.nasa.gov/Groups/SciTech/nano/
E-mail: jbluck@mail.arc.nasa.gov

The NAS Nanotechnology Team is interested in the application of molecular nanotechnology to aerospace systems. Their primary goals are to develop Ames' computational molecular nanotechnology capabilities; to design and computationally test atomically precise electronic, mechanical, and other nanoscale components; and to work with experimentalists to advance physical capabilities.

National Institute of Standards and Technology (NIST)

Department of Commerce
100 Bureau Drive

Gaithersburg, MD 20899-0001
(301) 975-NIST
URL: http://www.nist.gov/
E-mail: inquiries@nist.gov

NIST conducts and sponsors research in nanotechnology through its Electron Physics Group and other divisions. One focus of NIST research is the study of ultrasmall structures for the purpose of developing faster and smaller electronic devices. Scanning tunneling microscopes are used to manipulate individual atoms and molecules in this line of research.

Another major focus of NIST research is in the area of magnetic microstructures and nanostructures. The purpose of this research is to determine the behavior of electrons at the nanoscale to find ways of packing more electronic data into smaller spaces on computer hard drives and in thin-film memory devices.

National Institutes of Health (NIH)
Bethesda, MD 20892
URL: http://www.nih.gov/

NIH have become increasingly involved in the analysis of ways in which molecular nanotechnology may have implications for human health issues. In 1999, the U.S. Congress asked the National Institute of Environmental Health Sciences to explore the ways in which nanotechnology might be used to address environmental health problems. The study carried out in response to this request produced many possible applications.

One such application involves the use of nanoscale sensors to identify changes that take place within cells in response to toxins introduced into the cell. Another set of applications involves the construction of precisely ordered nanostructures that could be used to trap environmental and chemical pollutants. Still another possible use of nanotechnology involves the construction of nanoscale "scavengers" that could be used to remove substances from coal, oil, and other fuels that otherwise would result in the formation of air pollutants.

One of the concrete developments in the area of environmental health sciences so far has been the invention of a "ToxChip" by J. Carl Barrett and colleagues at the Laboratory of Molecular Carcinogenesis. The ToxChip contains thousands of modified DNA sequences on a glass chip. When a mixture of substances is passed over the chip, certain components in the mixture may bind to specific DNA sequences. This process permits the identification of possible toxic agents and an analysis of the mechanism by which they act in cells.

NIH also has sponsored a major conference on the application of nanotechnology to biomedical research. The conference was held on June 25 and 26, 2000, in Bethesda, Maryland, and was attended by researchers from many fields of science. Some conference topics at the conference were bioactive and biomimetic nanostructures for therapeutics and diagnostics, tissue engineering, the use of nano-structures in tissue repair, molecular motors, biosensors, targeted delivery of drugs, gene therapy, and early detection of disease.

National Science Foundation
4201 Wilson Boulevard
Arlington, VA 22230
Tel: (703) 306-1234/(800) 877-8339
URL: http://www.nsf.gov/

The National Science Foundation's mission is to promote the progress of science, engineering, and education in the United States. To that end, it has become increasingly active in the support of research in nanotechnology and the dissemination of information about the subject. In 1998, the Foundation announced a new Partnership in Nanotechnology: Synthesis, Processing, and Utilization of Functional Nanostructures (FNS).

A fundamental goal of FNS was to make possible team approaches to research in nanotechnology in which individuals from many subject matter areas were involved. That goal reflects one of the basic concerns about research in this area—that a variety of disciplines, ranging from physics and chemistry to computer science and biology, eventually will be involved in the design and fabrication of nanoscale devices.

FNS includes four "high-risk/high-gain" research areas: (1) synthesis and fabrication of nanostructures; (2) processing and conversion of molecules and nanoprecursors into functional nanostructures; (3) physical, mathematical, chemical, and biological modeling and simulation of nanoscale techniques; and (4) determination of fundamental physical, chemical, and biological properties of nanostructures and nanoscale interfaces.

Oak Ridge National Laboratory
Computational Nanotechnology
PO Box 2008
Oak Ridge, TN 37831
URL: http://www.ornl.gov/divisions/casd/polymer/02_nanotech.html

The Computational Nanotechnology project is involved with the modeling and simulation of components of nanomachines and with fundamental chemical physics at the nanometer scale. Some more recent projects include the modeling or simulation of the dynamics of a molecular bearing, nanometer-scale laser driven motors, and the dynamics of fluid flow in carbon nanotubes.

Paul Scherrer Institut (PSI)

CH-5232 Villigen PSI
Switzerland
Tel: +41-056-310-2111
Fax: +41-056-310-2199
URL: http://www.psi.ch/
E-mail: pr@psi.ch

PSI is an affiliate of the Swiss Federal Institute of Technology. It is a multidiscplinary center for research in the natural sciences and technology. Its primary areas of interest are solid-state physics, materials sciences, elementary particle physics, life sciences, nuclear and non-nuclear energy research, and energy-related ecology.

One of the research groups with PSI is the Laboratory for Micro- and Nanotechnology. Research at the laboratory now focuses on "top-down" and "bottom-up" approaches to nanotechnology. In the former field, some approaches being studied are electron-beam lithography, hot embossing lithography, focused ion beam science and technology, and injection molding. The focus of research in the area of molecular nanotechnology is the study of functional molecules on surfaces at micrometer and nanometer dimensions. Current projects in this area include research on biochemical recognition of individual molecules and nerve growth on microstructures designed for biological structures.

The RAND Corporation

1700 Main Street
PO Box 2138
Santa Monica, CA 90407-2138
Tel: (310) 393-0411
Fax: (310) 393-4818
URL: http://www.rand.org/
E-mail: correspondence@rand.org

The RAND Corporation was created after World War II for the purpose of studying national security issues. In the 1960s, it began to include domestic problems in its research program. Currently the center conducts studies in the areas of national defense, education and training, health care, criminal and civil justice, labor and population, science and technology, community development, international relations, and regional studies.

RAND's work is carried out in a variety of locations and settings, including the Council for Aid to Education, Institute for Civil Justice, RAND Education, RAND Enterprise Analysis, RAND Health, and Science and Technology. The company produces a large number and variety of publications, including RAND Reports, books, research briefs, issue papers, the *RAND Review*, and *The RAND Journal of Economics*. In 1995, RAND published an important report on the potential of nanotechnology for the development of molecular machinery (see Chapter 11 for details).

U.S. Department of Energy (DOE)
1000 Independence Avenue, SW
Washington, D.C. 20585
Tel: (202) 586-5000
URL: http://gils.doe.gov/

DOE has supported research in nanotechnology at many of its satellite laboratories and, through grants, at many academic laboratories. Some examples of this research include the following:

Surface Characterization and Control

One of the preliminary stages in nanotechnology research involves the process of understanding the characteristics and behavior of the upper layer of atoms or molecules in a material. This information is needed to determine how particles can be extracted, deposited, and moved about on the surface. DOE-supported research in this area is being conducted at the University of California at Berkeley's Advanced Light Source (ALS). The ALS is attached to the university's synchrotron-radiation particle accelerator, which produces photon beams in the 30 to 1,500 electron volt range. This beam permits the examination of the upper few layers of particles in a surface with precision.

Scanning Electron Microscope Studies

DOE also supports studies with scanning tunneling and atomic force microscopes used to move individual particles around on the surface of a material. Nestor Zaluzec at the Argonne National Laboratory outside Chicago is using a modified scanning electron microscope donated by Texas Instruments to characterize and modify the surface of semiconductor and other materials.

Development of Biomimetic Molecules

DOE's Sandia National Laboratories is engaged in research on the development of biomimetic molecules that may be useful in nano-technology syntheses. Biomimetic molecules are artificial molecules whose structures or functions or both are similar to those of molecules, such as proteins and nucleic acids, that occur naturally in living systems.

Nanoscale Catalysis

DOE has supported research at the Lawrence Berkeley Laboratory (LBL) in Berkeley, California, on nanocatalysis. Nanocatalysis is the process by which a scanning tunneling or atomic force microscope is used to bring about changes among atoms and molecules on the surface of a material. In 1995, researchers at LBL reported that they had used an atomic force microscope whose tip was coated with platinum to add hydrogen atoms to azide molecules on the surface of a sub-strate, converting the azide to an amine.

Industrial Organizations

Atomasoft Corporation
PO Box G9A 5H7
Trois-Rivières, QC 3351
Canada
Tel: (501) 227-9310
Fax: (775) 628-4628
URL: http://www.atomasoft.com/
E-mail: Info@Atomasoft.com

The company's name, derived from the words *ato*m, *ma*chine, and *soft*ware describes its function: the simulation and design of programs for prototype devices needed in the development of nanotechnology. The company provides engineering and consulting services and op-portunities for design, simulation, intellectual property, and manufac-turing partnerships.

California Molecular Electronics Corporation (CALMEC)
50 Airport Parkway
San Jose, CA 95110-1011

Tel: (408) 451-8404
Fax: 437-7777
URL: http://www.calmec.com
E-mail: jjmjr@calmec.com

CALMEC was founded in 1997 for the purpose of doing research and development in the field of molecular electronics. According to its mission statement, the company intends to "develop and sell quality products based on Molecular Electronics technology, sell and collect license fees on the rights to use its Molecular Electronics intellectual property, and build royalty streams on products developed and sold by others because of the application of its Molecular Electronics intellectual property." In addition to its Research and Development department, the company has a Business Development and Sales division responsible for marketing its new products and for providing technical support to its clients and customers.

Carbon Nanotechnologies, Inc.
16200 Park Row
Houston, TX 77084-5195
Tel: (713) 529-7264
Fax: (713) 529-7266
URL: http://carbonnanotech.com/
E-mail: marek@cnanotech.com

Carbon Nanotechnologies was founded in early 2000 by Richard Smalley, one of the discoverers of buckyballs, and his colleagues. The company describes itself as "the preeminent world player for [the manufacture of] single-wall carbon nanotubes."

Energenius Centre for Advanced Nanotechnology (ECAN)
c/o Department of Metallurgy and Materials Science
Wallberg Building, University of Toronto
184 College Street
Toronto, Ontario M5S 3E4
Canada
Tel: (416) 978-3012
URL: http://www.utoronto.ca/~ecan/

ECAN is dedicated to advancing research and training of students in the area of semiconductor nanotechnology. Some of the projects supported by the center include nanolithography, fabrication of novel quantum devices and their testing on a nanoscale, studies of hybrid nanodevices and circuits, and modeling of scanning probe microscope (SPM) tip interactions with semiconductor surfaces.

IBM Almaden Research Laboratory
650 Harry Road
San Jose, CA 95120-6099
Tel: (408) 927-1080
URL: http://www.almaden.ibm.com/
E-mail: info@almaden.ibm.com

The Almaden Research Center is one of three IBM sites (the others being the T. J. Watson Research Center and the Zurich Research Laboratory) at which significant research in molecular nanotechnology is being conducted. Included among the projects researchers are working on are the moving and positioning of single atoms and molecules, development and use of the magnetic force microscope and the magnetic-resonance force microscope, and research on sub-100 nm interferometric lithography.

IBM T. J. Watson Research Center
PO Box 218
Yorktown Heights, NY 10598
Tel: (914) 945-3000
Fax: (914) 945-2141
URL: http://www.research.ibm.com/nanoscience/
E-mail: avouris@us.ibm.com

Nanoscale research at this IBM center is carried out using scanning tunneling and atomic force microscopy and electron-beam lithography. Projects are aimed at the modification of materials at the atomic and molecular scale and the fabrication and study of nanoelectronic devices. Two important fields of research focus on the structures and properties of carbon nanotubes and their potential applications in nanoelectronic devices and nanometer-scale localized oxidation of semiconductors, with special attention to its potential application to the development of novel electronic devices.

IBM Zurich Research Laboratory (IBM-ZRL)
Säumerstrasse 4/Postfach
CH-8803 Rüschlikon
Switzerland
Tel: +41 1 724-8111
Fax: +41 1 724-8964
URL: http://www.zurich.ibm.com/
E-mail: info@zurich.ibm.com

Some of the most important advances in the field of molecular nanotechnology have come from IBM-ZRL. The scanning tunneling

microscope and the atomic force microscope were invented by workers associated with IBM-ZRL. Current research projects are aimed at
using these and other instruments to manipulate individual atoms and
molecules for the construction of nanoscale devices. One area of special interest is the development of nanoscale sensors for use in studying chemical and biological interactions, stress, and magnetization.

**Materials Research Science and Engineering Center (MRSEC)
  Program**
National Science Foundation
4201 Wilson Boulevard
Arlington, VA 22230
Tel: (703) 306-1234 (general NSF information); (703) 306-1996/ 1832/
  1814 (specific information on MRSEC)
E-mail: info@nsf.gov
URL: http://www.nsf.gov/

The MRSEC Program was established within the National Science
Foundation in 1994. The purpose of MRSEC is to provide funding
for research projects whose scope and complexity would make it inefficient for them to be funded individually by traditional means. Every
2 years, a certain number of academic centers are selected for funding
by the National Science Foundation under the program. As of early
2000, 27 universities had been designated as MRSEC locations.

MRSEC supports many fields of research, including dynamics, reactions, and catalysis of surfaces; structure materials, interfaces, grain
boundaries, and nanomechanics; polymeric materials and polymer
science; electronic and photonic materials; nanophase and nanostructured materials; mesoscopic systems; and biomolecular and biomi-
metric materials, self-assembly, and colloids.

Sites at which research on nanoscale devices, materials, and processes
now is being supported include:

- Center for Materials for Information Technology, University of
  Alabama
- Center for Advanced Materials Research, Brown University
- Materials Research Science and Engineering Center, Columbia
  University
- Materials Research Center, Harvard University
- Center for Nanostructured Materials, Johns Hopkins University
- Advanced Carbon Materials Center, University of Kentucky

- Center for Science and Engineering, Massachusetts Institute of Technology

Please see under Academic Organizations for more detail on the above-listed programs.

Molecular Electronics Corporation (MEC)
Two Prudential Plaza
180 North Stetson Avenue, Suite 850
Chicago, IL 60601
Tel: (312) 861-4300
URL: http://www.molecularelectronics.com/
E-mail: info@molecularelectronics.com

MEC was founded in December 1999 by Mark Reed, James Tour, and their associates for the purpose of developing commercial products based on their inventions and discoveries in the field of molecular electronics. Among the patents held by the company are those for spontaneously assembled molecular transistors and circuits, a programmable molecules device, a process for enzyme-catalyzed assembly, a low-power nonvolatile electronic memory device, and molecular electronic passive and active interconnect systems. The company's ultimate goal is to invent and develop electronic devices and information processes systems thousands of times smaller than existing devices and systems that consume less power and are far less expensive.

Nanogen
10398 Pacific Center Court
San Diego, CA 92121
Tel: (858) 410-4600
Fax: (858) 410-4848
URL: http://www.nanogen.com/
E-mail: prodinfo@nanogen.com

Nanogen was founded in 1993 for the purpose of developing the technology needed to integrate microelectronics and molecular biology on semiconductor microchips. The technology used by the company is based on the fact that most biological molecules are electrically charged. They can be attached to semiconductor microchips in any variety of patterns needed for a specific application. Unknown materials added to such a microchip can be identified according to the way in which they bind to the template.

Nanogen expects that its technology will have applications in a variety of fields, such as biomedical research, medical diagnostics, genomics research, genetic testing, and drug development. Future applications are anticipated in the fields of agricultural, environmental, and industrial analyses.

Nanophase Industries Corporation

453 Commerce Street
Burr Ridge, IL 60521
Tel: (630) 323-1200
Fax: (630) 323-1221
URL: http://www.nanophase.com/
E-mail: info-w@nanophase.com

Nanophase Industries manufactures and sells nanocrystalline materials specially formulated for a variety of applications. The company uses its own patented physical vapor synthesis process for producing these materials. In this process, a solid material is vaporized by a burst of energy. The particles formed in this process are treated with a reactive gas and condensed. The particles contain anywhere from a few thousand to a few tens of thousands of atoms.

Nanocrystalline particles formed in this way have optical, electrical, magnetic, and chemical properties different from those of larger size. They now are being used in applications such as abrasives, catalysts, cosmetics, electronic devices, magnetics, pigments and coatings, and structural ceramics.

NanoPowders Industries

23 Hata'as Street, Suite 300
PO Box 2368
Kfar Saba, Israel 44425
Tel: +972-9-7663880
Fax: +972-9-7663878
or
1111 Kane Concourse, Suite 400
Bay Harbor Islands, FL 33154
Tel: (305) 861-0818
Fax: (305) 861-1663
URL: http://www.nanopowders.com/
E-mail: webmaster@nanopowders.com

NanoPowders Industries was founded in 1994 to develop, manufacture, and market nanoscale powders made from precious and base metals. The company currently offers more than 50 different formulations from a variety of metals, such as silver, palladium, aluminum, nickel, and copper.

The dimensions of particles available in these formulations is of the order of a few nanometers, at least 20 times smaller than the average size of metal powders currently available. These particles have electrical and mechanical properties significantly different from those larger in size. The company anticipates that its products will find application primarily in four fields: electronics, medical, batteries, and industrial catalysts.

Nanosystems Research Group
The MITRE Corporation
202 Burlington Road
Bedford, MA 01730-1420
Tel: (781) 271-2000
and
1820 Dolly Madison Boulevard
McLean, VA 22102-3481
Tel: (703) 883-6000
URL: http://www.mitre.org

The Nanosystems Research Group at the MITRE Corporation pursues a variety of research projects dealing with nanoscale devices and systems. Its primary focus is on the modeling of nanosystems that will result in virtual prototypes of nanodevices and nanocomputers. Some of the specific tasks include research on nanocomputer technologies, self-assembly of nanoscale and microscale systems, and economic analysis of nanotechnology.

The Nanosystems Research Group's website contains a collection of some of the most complete and interesting information currently available about nanotechnology. It includes "top-down" and "bottom-up" approaches to the subject. See the listing for the "Nano-electronics and Nanocomputing Home Page" in Chapter 12 for more detail.

Technanogy
Irvine Commercialization Center
1601 Alton Parkway, Suite B
Irvine, CA 92606
Tel: (949) 261-1420
Fax: (949) 261-0311
http://www.technanogy.net
E-mail: info@technanogy.net

The primary goal of Technanogy is to find ways of converting the discoveries made in nanoscale science and engineering into successful commercial products and ventures. The company invests capital, expertise, and intellectual property into a network of related groups dedicated to the discoveries and inventions themselves and to their ultimate commercial realization. The company was founded in 1999 as Nanopropulsion with the aim of finding ways of improving the efficiency of rocket motors. Its rapid success in this endeavor convinced the company's founders, Larry Welch and Joe Martin, that many more nanoscale discoveries were on the horizon, and they decided to expand the focus of the company to the broader horizons now encompassed by Technanogy.

Xerox Palo Alto Research Center (Xerox PARC)
3333 Coyote Hill Road
Palo Alto, Ca 94304
Tel: (650) 812-4000
URL: http://www.parc.xerox.com/
E-mail: webmaster@parc.xerox.com

PARC was established by the Xerox Corporation in 1970 with the goal of creating "the architecture of information." To carry out this mission, the company brought some of the best researchers available to PARC. Since that time, many important discoveries and inventions have been produced at PARC. Among these are The Alto, the first personal computer; Ethernet, a system for interconnecting computers in local area networks; glyphs, systems for transforming paper data into computer records; flat panel displays for televisions and computer screens; laser diodes; and multibeam and blue lasers. For the decade during which Ralph Merkle was employed at Xerox PARC, the center was also the source of some of the most imaginative theoretical research and computer modeling of devices, systems, and procedures for use in molecular nanotechnology.

Zyvex LLC
1321 North Plano Road, Suite 200
Richardson, TX 75081
Tel: (972) 235-7881
Fax: (972) 235-7882
URL: http://www.zyvex.com/
E-mail: infor@zyvex.com

Zyvex was founded in 1997 by Jim Von Ehr as the first company created for the primary purpose of developing molecular nanotechnology. Its goal is to build the an assembler, one of the fundamental devices required for progress in molecular nanotechnology.

Zyvex has announced that it is currently funded at a level sufficient to carry out its own research and to sponsor research projects at universities. It anticipates receiving no revenues for its work until sometime between 2005 and 2010.

Academic Organizations

Also see list of courses in nanoscale science and engineering at the end of this section.

Advanced Carbon Materials Center (ACMC)

University of Kentucky
27 Chemistry-Physics Building
Lexington, KY 40506-0046
Tel: (606) 257-8844
Fax: (606) 323-1069
URL: http://www.mrsec.uky.edu/
E-mail: haddon@pop.uky.edu

ACMC is a National Science Foundation–sponsored Materials Research Science and Engineering Center Program. It is a division of the university's Center for Applied Energy Research. The primary focus of research at ACMC is the low-temperature synthesis of single-wall carbon nanotubes and the study of possible commercial applications for devices and materials made from these structures.

Beckman Institute for Advanced Science and Technology

Office of External Relations
University of Illinois at Urbana–Champaign
405 North Matthews Avenue
Urbana, IL 61801
Tel: (217) 244-5582
Fax: (217) 244-8371
URL: http://www.beckman.uiuc.edu/
E-mail: jjones@director.beckman.uiuc.edu

The Beckman Institute has been one of the nation's premier research centers since it opened in 1989. It traditionally has recognized the importance of interdisciplinary research and is following that practice in its current nanoscale research. Research at the institute currently is divided into three main areas: Biological Intelligence, Human-Computer Intelligent Interaction, and Molecular and Electronic Nanostructures. Some of the specific topics being explored include biomolecular electronics, computer modeling of nanoscale devices, and protein engineering. Researchers also are using scanning tunneling microscopes to manipulate the atoms on the surface of clean materials and are studying the design and construction of self-assembly nanostructures.

California NanoSystems Institute (CNSI)

Department of Chemistry and Biochemistry
University of California at Los Angeles
Box 951569
Los Angeles, CA 90095-1569
Tel: (310) 825-2836
URL: http://www.cnsi.ucla.edu/
E-mail: heath@chem.ucla.edu

California's Governor Gray Davis announced in December 2000 that CNSI had been selected as one of three statewide research programs chosen to receive $100 million in state support of scientific research. CNSI is a joint venture involving the University of California at Los Angeles and the University of California at Santa Barbara. Three major foci of research at CNSI will be on (1) the development of chemical, biological, and materials approaches toward the fabrication of nanoscale building blocks of macroscopic materials and devices; (2) the characterization of tools for understanding the structure and properties of materials at the nanoscale level; and (3) the development of manufacturing tools for the construction of nanosystems. Two general topics of research within the institute will be Nanosystems: Molecular Medicine and Nanosystems: Information Technology.

Center for Advanced Materials Research (CAMR)

Brown University, Box M
Providence, RI 02912
Tel: (401) 863-2184
Fax: (401) 863-1387
URL: http://www.brown.edu/Departments/Advanced_Materials_
   Research/CAMR/
E-mail: CAMR@brown.edu

CAMR is an independent academic unit at Brown University, sponsored and funded by the National Science Foundation's Materials Research Science and Engineering Center Program. The focus of CAMR's research is on the micromechanics and nanomechanics of materials, with a view toward improving understanding of the mechanisms of deformation and failure in solid materials. One consequence of this research should be the development of new and stronger materials with applications in both structural and electronic devices.

Center for Atomic-Scale Materials Physics (CAMP)
Department of Physics
Technical University of Denmark
Building 307-309
DK-2800 Lyngby
Denmark
Tel: +45 4588 1611
URL: http://www.fysik.dtu.dk/CAMP/
E-mail: Webmaster@fysik.dtu.dk

CAMP was founded by the Danish National Research Foundation in 1993. It maintains experimental and theoretical activities at the Technical University of Denmark and the University of Aarhus. Some current projects at CAMP are anchoring of organic molecules to a metal surface, atomic-scale structure of Co-Mo-S nanoclusters, nanoclusters in HDS (hydrodesulfurization) catalysis, and point contacts and nanowires.

Center for Biological and Environmental Nanotechnology (CBEN)
Rice University
PO Box 1892
Houston, TX 77251-1892
Tel: (713) 348-8210
Fax: (713) 348-8218
URL: http://cnst.rice.edu/
E-mail: sdcarpen@rice.edu

The Center for Biological and Environmental Nanotechnology was established in 2001 as one of the National Science Foundation's Nanoscale Science and Engineering Centers. Its work has three primary foci: (1) research on the "wet/dry" interface, that is, the interaction between aqueous solutions and nanoscale materials and devices; (2) programs of education for 9th grade students in the Houston school district, and (3) development of partnerships between academia and industry with the goal of accelerating the transition from nanoscience to nanotechnology research.

Center for Electronic Transport in Molecular Nanostructures
Columbia University
2960 Broadway

New York, NY 10027-6902
Tel: (212) 854-1754
Fax: (212) 749-0397
URL: http://www.cise.columbia.edu/nsec/
E-mail: beb2@columbia.edu

The Center for Electronic Transport in Molecular Nanostructures was chosen in late 2001 as one of the National Science Foundation's Nanoscale Science and Engineering Centers. Its primary emphasis will be on the development of an understanding of the way electrical charges are transported through nanoscale materials and devices, including new materials that may be developed at the center. The center will also analyze possible applications of its research to a variety of scientific fields, including photonics, biology, neuroscience, and medicine.

Center for Materials for Information Technology (MINT)

The University of Alabama
Box 870209
Tuscaloosa, AL 35487-0209
Tel: (205) 348-2516
Fax: (205) 348-2346
URL: http://bama.ua.edu/~mint/
E-mail: dsnow@mint.ua.edu

MINT is sponsored and funded by the National Science Foundation's Materials Research Science and Engineering Center Program. It consists of 22 faculty members from seven academic departments at the university. The major emphasis of the MINT program is the development of new materials for advanced data storage. More recent research has dealt with electrical, magnetic, and optical properties of thin films and nanoparticles and nanostructures.

Center for Materials Science and Engineering (CMSE)

Massachusetts Institute of Technology (MIT)
77 Massachusetts Avenue, Room 13-2106
Cambridge, MA 02139-4307
Tel: (617) 253-6850
Fax: (617) 258-6478
URL: http://web.mit.edu/cmse/
E-mail: cmse-www@mit.edu

CMSE at MIT is part of the National Science Foundation's network of Materials Research Science and Engineering Centers. Its staff consists of 100 faculty members from 11 MIT academic departments. Research at CMSE is carried out in five Interdisciplinary Research

Groups, focused on (1) Microphotonic Materials and Structures, (2) Nanostructured Polymers, (3) Electronic Transport in Mesoscopic Semiconductor Structures, (4) Microstructure and Mechanical Properties of Polymeric Materials, and (5) Doped Mott Insulators.

Center for Nanoscience and Technology

Department of Electrical Engineering
University of Notre Dame
Notre Dame, IN 46556
Tel: (219) 631-8673
Fax: (219) 631-4393
http://www.nd.edu/~ndnano/
E-mail: Ndnano@nd.edu

Research at the center is focused on multidisciplinary investigation of concepts in nanoscience and engineering with emphasis on applications to unique functional capabilities. Some topics of current interest are quantum cellular automata and architectures; resonant tunneling devices and circuits; and photonic high-speed nano-based materials, devices, and circuits.

Center for Nanostructured Materials and Interfaces

4639 Engineering Hall
1415 Engineering Drive
University of Wisconsin at Madison
Madison, WI 53706
Tel: (608) 263-2922
Fax: (608) 265-3782
URL: http://mrsec.wisc.edu\
E-mail: kuech@engr.wisc.edu

The center is organized into three interdisciplinary research groups (IRGs) and four smaller seed groups studying a variety of nanoscale issues. IRG1 is working on issues in materials integration on silicon. IRG2 is researching the role of grain boundaries and electronic structures in high-temperature superconductors. IRG3 is studying the design, synthesis, and processing of three-dimensional structures with controlled topography on the 1- to 100-nm scale.

Center for Nanotechnology

NANOTECH USER FACILITY
University of Washington
Box 352140
Seattle, WA 98195

Tel: (206) 616-2118
URL: http://www.nano.washington.edu/
E-mail: dqin@u.washington.edu

The Center for Nanotechnology was created in 1997 to study many problems related to nanoscale science and engineering. Included among these problems are an improved understanding of the workings of natural bionanosystems, developing new nanoprocesses in medicine, obtaining a better understanding of the properties of single molecules, and creating modeling and simulation systems for nanoscale objects and events.

Center for Research at the Bio/Nano Interface
Chemistry Department
University of Florida
PO Box 117200
Gainesville, FL 32611-7200
Tel: (352) 392-0541
Fax: (352) 392-8758
URL: http://www.chem.ufl.edu/Divisions/Analytical/bio_nano/
E-mail: crmartin@chem.ufl.edu

The Center for Research at the Bio/Nano Interface involves researchers from many departments at the University of Florida who are interested in the principles and applications of biochemical, biological, and biomedical processes at the microscale and nanoscale. The center promotes collaborative research, acts as a clearinghouse for information, and provides a forum for the writing of joint proposals. Some research topics now being pursued at the center include research on protein channels and other biochemical microstructures, applications of nanoparticles in biomedicine, and research on single-molecule or biomolecule detection.

Centre for Nano-Device Modelling (CNDM)
Department of Applied Mathematics
University of Leeds
Woodhouse Lane
Leeds LS2 9JT
UK
Tel: +44 (0) 113 233-5110
Fax: +44 (0) 113 242-9925
URL: http://amsta.leeds.ac.uk/cndm/
E-mail: amt6eac@amsta.leeds.ac.uk/

CNDM coordinates the work of four university departments engaged in the modeling of nanoscale devices, including the Department of Applied Mathematics, the School of Electronic and Electrical Engineering, the Department of Physics and Astronomy, and the Centre for Self-Organising Molecular Systems. Details of the research being conducted in each of these departments and schools can be found on their respective webpages, listed on the CNDM page. Much of the work being conducted in various departments involves collaboration with other departments involved in the CNDM, and new joint projects always are under consideration.

Centre for Nanoscale Science & Technology
University of Newcastle upon Tyne
Newcastle upon Tyne NE1 7RU
UK
Tel: +44 (0) 191 222 7362
Fax: +44 (0) 191 222 7361
URL: http://nanocentre.ncl.ac.uk/m21.htm
E-mail: CNSAT@ncl.ac.uk

The Centre for Nanoscale Science & Technology was established in 1999 to pursue research and training programs in microscale, nanoscale, and molecular-scale fabrication of new materials for possible applications in genetic diagnostics, drug discovery, and chemical and environmental monitoring. Some specific areas of interest include biological, chemical, and drug sensors; nanostructure fabrication and synthesis; nanoelectronic and microelectronic devices; and instrumentation for fabrication and characterization.

Center for Science in Nanometer Scale
Seoul National University
Shillim-dong, Kwanak-ku
Seoul, 151-742
Republic of Korea
Tel: +82 2 876-9926 or +82 2 880-8644
Fax: +82 2 887-6575 or +82 2 873-7039
URL: http://csns.snu.ac.kr/
E-mail: netty@isrca.snu.ac.kr

The Center for Science in Nanometer Scale was established in October 1997 to conduct research on materials in the nanoscale with particular attention to applications in memory devices.

Center for Science of Nanoscale Systems and their Device Applications
Harvard University
c/o Dr. Robert M. Westervelt
Pierce 234/McKay 203
Cambridge, MA 02138
Tel: (617) 495-3296
Fax: (617) 495-9837
URL: http://www.nsec.harvard.edu
E-mail: westervelt@deas.harvard.edu

The Center for Science of Nanoscale Systems and their Device Applications was established in 2001 as part of the National Science Foundation's Nanoscale Science & Engineering Centers. The center is a collaborative effort involving, in addition to Harvard, the Massachusetts Institute of Technology; the University of California at Santa Barbara; the Museum of Science in Boston; the Brookhaven, Oak Ridge, and Sandia National Laboratories; and the Delft University of Technology in The Netherlands. The focus of research at the center will be on a combination of "top-down" and "bottom-up" technologies that will produce novel electronic and magnetic devices.

Centre for Self-Organizing Molecular Systems (SOMS)
University of Leeds
Leeds LS2 9JT
UK
Tel: +44 (0) 113 233 6453
Fax: +44 (0) 113 233 6452
URL: http://chem.leeds.ac.uk/SOMS/
E-mail: soms@chem.ac.uk

SOMS is an interdisciplinary program established for the purpose of learning more about biological molecular self-assembly systems and the application of such systems for the development of new materials with applications in electronics, sensors, and medicine. SOMS applies an interdisciplinary approach to its projects, drawing on the resources of biology, chemistry, physics, and engineering.

In describing its efforts, SOMS argues that "molecular nanotechnology is perceived to be an inevitable development, albeit very long term." In this context, the goal of SOMS is to "become increasingly aligned to the needs of nanotechnology." As an example of this work, one project focuses on the design of "novel, hybrid, nanoscale device architectures" with potential applications in information storage devices, advanced gas sensors, and the imaging of large bio-molecules.

Department of Applied Physics
Delft University of Technology
Lorentzweg 1
2628 CJ Delft
PO Box 5046
2600 GA Delft
The Netherlands
Tel: +31 15 278 4841
Fax: +31 15 278 2655
URL: http://www.tn.tudelft.nl/INDEX.HTM
E-mail: info@tnw.tudelft.nl

Some of the most exciting developments in molecular nanotech-nology have come from the laboratory of Cees Dekker, at the Delft University of Technology. The Department of Applied Physics carries out research in many areas of nanotechnology, with "top-down" and "bottom-up" approaches. Some examples include studies on molecu-lar quantum wires and quantum dots, single-electron tunneling devices in silicon, nanostructures for two-dimensional photonic waveguides, and the use of scanning probe microscopy for studies in nanotribology.

Institute for Nanotechnology
Northwestern University
2145 Sheridan Road, TECH K111
Evanston, IL 60208
Tel: (847) 491-5784
Fax: (847) 491-3721
URL: http://www.nsec.northwestern.edu/
E-mail: nanotechnology@northwestern.edu

The Institute for Nanotechnology consists of three major research centers: (1) the Center for Nanofabrication and Molecular Self-Assembly, (2) the Nanoscale Science and Engineering Center (NSEC), and (3) the Center for Transportation Nanotechnology. All three cen-ters are interdisciplinary in nature, drawing on resources from acade-mia, industry, and national laboratories. NSEC is also home to the Center for Integrated Nanopatterning and Detection Technologies (CINDT), one of the National Science Foundation's Nanoscale Sci-ence and Engineering Centers. Research at CINDT will focus on the creation of new structures, molecule by molecule (nanopatterning), the development of materials that can sense and bind to a variety of chemical and biological agents, and the synthesis of devices that will emit a detectable optical or electrical signal when such binding occurs.

Interactive Nano-Visualization in Science and Engineering Education (IN-VSEE) Project

Center for Solid State Science
Arizona State University
Tempe, AZ 85287
Tel: (480) 965-0919
Fax: (480) 965-1500
URL: http://invsee.asu.edu/Invsee/
E-mail: invsee@asu.edu

IN-VSEE is a project funded under the National Science Foundation's Applications of Advanced Technologies program. It is a consortium of university and industry scientists and engineers, college and high school science teachers, and museum educators. The goal of the project is to develop an educational program that will help students be prepared better for the coming revolution in nanotechnology.

The project has developed a series of interactive modules in which students are able to operate advanced microscopes and nanofabrication tools on the Internet. The modules included in the program are Size and Scale, Scanning Probe Microscopy, The Music of Spheres, The Allotropes of Carbon, Why Does a Light Bulb Burn Out?, Modern Information Storage Media, The Five Kingdoms of Biology, What is Friction?, What is That in Your Dog Dish?, Iridescence, Biominerals, Biological Structural Materials, Engineered Materials, The World of Liquid Crystals, DNA—Infinite Variety in Such Small Packets, The Morphology and Use of Gold Films, and Osmotic Pressure in Red Blood Cells and Plant Cells.

Laboratory for Molecular Robotics

Computer Science Department
University of Southern California
Los Angeles, CA 90089-0781
Tel: (213) 740-4502
Fax: (213) 740-7512
E-mail: lmr@lipari.usc.edu

The Laboratory for Molecular Robotics was established in 1994 by Ari Requicha, who also taught the first regular university course based on Eric Drexler's book *Nanosystems*. Funding is provided by the Zohrab A. Kaprielian Technology Innovation Fund. Early research at the laboratory has been making use of the scanning probe microscope to study the atomic characteristics of surfaces and to manipulate individual atoms and groups of atoms on those surfaces. The laboratory is an interdisciplinary program with participation from the Departments of Computer Science and Electrical Engineering, Materials Sciences and Physics, and the Information Sciences Institute.

Laboratory for Nanotechnology
657 S. Mechanic Street
Pendleton, SC 29670
Tel: (864) 646-9501
URL: http://virtual.clemson.edu/goups/nanotech/
E-mail: dcarrol@clemson.edu

The Laboratory for Nanotechnology maintains a variety of programs in physics, organized around certain focus groups. These focus groups currently include Nanostructure/Self-Assembly and Macromolecular Electronics, Organic Devices and Optics, and Transition-Metal-Oxide Studies. Laboratory Director David L. Carroll currently is studying nanomanipulation and modification of single-walled and multiwalled carbon nanotubes.

Materials and Process Simulation Center
Chemistry 139074
California Institute of Technology
Pasadena, CA 91125
Tel: (626) 395-2730
URL: http://www.wag.caltech.edu/
E-mail: wag@wag.caltech.edu

The Materials and Process Simulation Center at Caltech focuses on the *de novo* (or "first principles") design of industrial catalysts, drugs, nanoscale materials, and processes by using a combination of quantum, atomic-scale, mesoscale, and macroscopic techniques. Researchers are attempting to develop new theories, methods, and software to describe the chemical, biological, and materials properties of systems, then to validate these methods and apply them to actual problems faced by engineers. The Caltech team was awarded the 1999 Feynman Molecular Nanotechnology Theory Prize for its research on nanoscale materials.

Materials Research Science and Engineering Center (MRSEC)
Columbia Radiation Laboratory
530 W. 120th Street

Room 1001 Schapiro CEPSR
Mail Code 8903
Columbia University
New York, NY 10027
Tel: (212) 854-3964
Fax: (212) 854-1909 (Attn: MRSEC)
URL: http://research.radlab.columbia.edu/mrsec/
E-mail: ipbl@columbia.edu

MRSEC was established in 1998 with a grant from the National Science Foundation as part of its Materials Research Science and Engineering Center Program. The Columbia MRSEC consists of a single multidisciplinary group consisting of faculty from Columbia and other New York City academic institutions and researchers from industrial laboratories in the metropolitan New York City area. The focus of research within this group is on the synthesis of nanoparticles; self-organization of these particles into useful films; and the electrical, optical, and other properties of these films and other aggregates.

Materials Research Science and Engineering Center (MRSEC)
Harvard University
208 Pierce Hall
29 Oxford Street
Cambridge, MA 02138
Tel: (617) 495-3760
Fax: (617) 496-4654
URL: http://www.mrsec.harvard.edu/
E-mail: mrsec@harvard.edu

The Harvard MRSEC was established in 1994 as part of the National Science Foundation's Materials Research Science and Engineering Center Program. It is an interdisciplinary project involving faculty members from the Departments of Chemistry, Earth and Planetary Sciences, and Physics and the Division of Engineering and Applied Sciences. Four Interdisciplinary Research Groups work under the auspices of MRSEC: (1) New Materials: Synthesis, Characterization and Theory; (2) Interfaces: Structure, Wetting and Motion; (3) Electronic and Photonic Nanostructures; and (4) Design and Manufacturing Initiative. Some topics of special interest in the area of nanotechnology include droplet nucleation on surfaces coated with self-assembled monolayers, nanoscale superconducting tunneling transistors, and fabrication of optical wave guides by microprinting and self-assembly.

Materials Research Science and Engineering Center (MRSEC)
The Johns Hopkins University
307 Bloomberg
3400 North Charles Street
Baltimore, MD 21218
Tel: (410) 516-8092
Fax: (410) 516-7239
URL: http://www.pha.jhu.edu/groups/mrsec/
E-mail: clc@pha.jhu.edu

MRSEC at Johns Hopkins is a project of the National Science Foundation's Materials Research Science and Engineering Center Program. It was established in January 1997 and focuses on the development and study of new nanoscale structures with zero, one, and two dimensions. Research deals with five aspects of this problem: (1) synthesis and processing, (2) measurement of properties, (3) theoretical modeling, (4) fabrication of prototype devices, and (5) applications. Examples of current research include studies of nanoreinforced thin films, reactive multilayer foils, nanocrystalline iron thin films, multilayered nanowires, semimetallic bismuth nanowires, and oxide nanoparticles.

Nano & Micromechanics Laboratory
Room 743, Barus and Holley Building
Brown University
Providence, RI 02912
Tel: (401) 863-1456
URL: http://en732c.engin.brown.edu/
E-mail: kim@engin.brown.edu

The Nano & Micromechanical Laboratory was created as the Micromechanical Laboratory in 1989 and was expanded to include research at the nanoscale level a year later. Research and instruction are focused on the experimental, computational, and conceptual study of nanomechanics and micromechanics of materials.

Nanobiotechnology Center
101 Biotechnology Building
Cornell University
Ithaca, NY 14853
Tel: (607) 254-5393
Fax: (607) 254-5375
URL: http://www.nbtc.cornell.edu/
E-mail: hgc1@cornell.edu/

The Nanobiotechnology Center is a consortium of Cornell and Princeton Universities, the Oregon Health Science University, and the Wadsworth Center of the New York State Department of Health. It was established in 1999 with a 5-year, $19 million grant from the National Science Foundation. The Center is a highly interdisciplinary institution whose goal is to increase understanding of the way biological systems function and to discover ways in which that new knowledge can be applied to the development of new nanoscale devices. The six research areas included in the Center's work include (1) microanalysis of biomolecules; (2) molecular templates; (3) bioselective surfaces; (4) selective molecular filtration; (5) sparse cell isolation; and (6) powering nanomachines with molecular motors.

Nanostructure Optoelectronics
School of Electrical and Computer Engineering
Georgia Institute of Technology
Atlanta, GA 30332-0250
Tel: (404) 894-2902
Fax: (404) 894-4641
URL: http://www.ece.gatech.edu/research/nanostructure_optoelectronics/research.html/
E-mail: guthrie@wilma.physics.gatech.edu

This research group was organized to explore the development of next-generation electronic devices at submicron dimensions. Examples of research now being conducted include work on nanostructure fabrication and characterization, quantum well lasers and light-emitting diodes, ballistic electron emission microscopy and spectroscopy, and diffractive and waveguide optics.

The Center of Nanostructure Materials and Quantum Device Fabrication (NanoFAB Laboratory)
College of Engineering
University of Texas at Arlington
Box 19019
Arlington, TX 76019-0019
Tel: (817) 272-2571
Fax: (817) 272-2548
URL: http://www-ee.uta.edu:80/NanoFab/
E-mail: info@engineering.uta.edu

The NanoFab Center was established at Texas A&M in 1990 to study nanoscale structures and develop nanotechnologies, especially nanoelectronics. It now operates as a consortium of researchers from the University of Texas at Dallas, the University of Texas at Arlington, and the Texas Engineering Experiment Station at College Station. Research at the center focuses on the fabrication and study of nanoscale structures, the development of new nanoscale technologies, and the distribution of information about nanoscale science and technology to the scientific community.

Nanomanipulator Project
Department of Computer Science
University of North Carolina at Chapel Hill
Chapel Hill, NC 27514
Tel: (919) 962-3526
URL: http://www.cs.unc.edu/Research/nano/index.html
E-mail: guthold@cs.unc.edu

The Nanomanipulator Project is a joint endeavor of the Departments of Computer Science, Physics & Astronomy, Chemistry, and Psychology; the Schools of Education and of Information and Library Science; and the Gene Therapy Group of the University of North Carolina at Chapel Hill and the North Carolina State Department of Computer Science. The program provides an avenue by which scientists and students in the areas of biology, solid-state physics, and gene therapy can have remote access to the University's scanning probe microscope in their research.

Nanomechanics and Tribology
Institute de Génie Atomique
Départment de Physique
Ecole Polytechnique Fédérale de Lausanne
1015 Lausanne
Switzerland
Tel: +41 21 693 3374
Fax: +41 21 693 4470
URL: wysiwyg://24/http://igahpse.epfl.ch/
E-mail: Willy.Benoit@epfl.ch

This research group takes a "bottom-up" approach to research on the elastic and anelastic properties of materials using acoustic and scanning probe microscopy. Some more recent research projects of the laboratory have dealt with nanomechanics; mechanical properties of carbon nanotubes; elastic and shear moduli of carbon nanotube ropes; and nanosubharmonics, the dynamics of small nonlinear contacts.

NanoStructures Laboratory
Massachusetts Institute of Technology
Room 39-427
Cambridge, MA 02139
Tel: (617) 253-7545
Fax: (617) 253-8509
URL: http://nanoweb.mit.edu/\/

The Nanostructures Laboratory at the Massachusetts Institute of Technology employs "bottom-down" techniques for the fabrication of nanoscale structures. Some of the processes used include scanning electron-beam lithography, x-ray nanolithography, interferometric lithography, and deep-ultraviolet contact photolithography. These procedures make use of various forms of electromagnetic radiation to etch designs into a substrate, with the challenge being to produce ever smaller and more precise lines and shapes.

NanoStructures Laboratory
Princeton University
School of Engineering and Applied Science
B-210 E-Quad
Princeton, NJ 08544-5263
URL: http://www.ee.princeton/edu/~chouweb/

Research at the NanoStructures Laboratory at Princeton is aimed at producing structures that are smaller, better, and less expensive than those currently available. The work of the laboratory is divided into five major categories: Nanofabrication, Nano-electronics, Nano-optoelectronics, Nano-magnetics, and Application of Nanostructures in Other Fields (such as semiconductors and polymers). The methods used in this research tend to be traditional "top-down" approaches, such as various types of lithography.

Nanostructures Research Group
Engineering Research Center
Arizona State University
CSSER/CEAS
Box 876206
Tempe, AZ 85287-6206
Tel: (480) 965-3708
Fax: (480) 965-8118
URL: http://www.eas.asu.edu/~nano/
E-mail: cssermail@asu.edu

The Nanostructures Research Group is an interdisciplinary group of researchers working within the College of Engineering's Center for Solid State Electronics Research. The focus of the group's work is on ultrasmall semiconductor devices, particularly on the modeling of such devices and the use of nanolithography for their production and study. Some general areas of study include quantum dots in silicon, physics of quantum transport, fast photoexcitation of semiconductors, and the development of processing technology for ultrasubmicron devices.

Cornell Nanofabrication Facility
Knight Laboratory, Cornell University
Ithaca, NY 14853-5403
Tel: (607) 255-2329
Fax: (607) 255-8601
URL: http://www.cnf.cornell.edu/
E-mail: webmaster@cnf.cornell.edu

Materials Science Research Center of Excellence
Howard University School of Engineering
2300 Sixth Street, NW
Washington, D.C. 20059
Tel: (202) 806-6618
Fax: (202) 806-5367
URL: http://www.msrce.howard.edu/~nanonet/NNUN.HTM
E-mail: nanonet@msce.howard.edu

Electronic Materials & Processing Research Laboratory
E. E. West Building
The Pennsylvania State University
University Park, PA 16802
Tel: (814) 865-0870
Fax: (814) 865-6581
URL: http://www.emprl.psu.edu/
E-mail: info@emprl.psu.edu

Stanford Nanofabrication Facility
Stanford University
CIS Building
Stanford, CA 94305-4070
Tel: (630) 725-3609
Fax: (630) 725-4659
URL: http://www-snf.stanford.edu/
E-mail: webmaster@snf.stanford.edu

Nanotech at UC Santa Barbara

Nanotech
Electrical and Computer Engineering
University of California
Santa Barbara, CA 93106
Tel: (805) 893-4130
Fax: (805) 893-8170
URL: http://www.nanotech.ucsb.edu
E-mail: nanotech@ece.ucsb.edu

National Nanofabrication Users Network (NNUN)

NNUN consists of research departments at the five above-listed U.S. universities with state-of-the-art equipment and facilities for research at the nanometer scale. Interested researchers from any university, government agency, or industrial corporation may apply to carry out their research at any one of the five institutions. In addition to equipment provided, the institutions have staffs with experience and expertise in nanoscale experimentation. The type of equipment generally available includes atomic force and scanning electron microscopes, spectrophotometers, nanoprobes, electron-beam lithography and photolithography equipment, and thin film deposition and etching equipment.

Research at the five NNUN sites covers a wide variety of disciplines, including superconducting materials, optical microscopy, laser etching of semiconductors, growth and properties of thin oxides, and production and characterization of nanostructures. Researchers come from many different fields, including pure and applied physics, computer sciences, electrical engineering, material sciences, and life sciences. A major focus of research at NNUN universities is the area of thin film deposition and characterization.

North Carolina Center for Nanoscale Materials (NCCNM)

Department of Physics and Astronomy
University of North Carolina at Chapel Hill
CB 3255 Phillips Hall
Chapel Hill, NC 27599
Tel: (919) 962-3297
Fax: (919) 962-0480
URL: http://www.physics.unc.edu/~zhou/muri
E-mail: zhou@physics.unc.edu

Research at NCCNM is focused on the experimental and theoretical study of the properties and potential applications of carbon nanotubes. The four general areas of research are new materials synthesis and fabrication, mechanical properties, electronic properties, and applications.

Project Nanostructures
Institut für Festkörperphysik
PN 5-2, Hardenbergstrasse 36
10623 Berlin
Germany
Tel: +49 30 314 22082
Fax: +49 30 314 22569
URL: http://sol.physic.TU-Berlin.DE/htm_group/

The primary focus of research in Project Nanostructures is on the fabrication, characterization, and theoretical modeling of quantum wires and quantum dots. Some specific research topics include the growth of self-organized nanostructures, mechanisms of self-ordering in quantum dot arrays, x-ray diffraction from ultrasmall nanostructures, and processing and characterization of quantum wire and quantum dot lasers.

Purdue Nanotechnology Initiative
Department of Physics
West Lafayette, IN 47907
URL: http://molecule.ecn.purdue.edu/~cluster/initiative.html
E-mail: ronald@ecn.purdue.edu

The Purdue Nanotechnology Initiative consists of researchers from the Departments of Chemical and Electrical Engineering, Chemistry, and Physics at Purdue University. The goal of the Initiative is to design and fabricate nanoelectrical devices consisting of metallic clusters joined by nanoscale wires. The structures produced as a result of this research have been named *linked cluster networks (LCNs)*. Members of the Initiative currently are studying the electrical, chemical, optical, and magnetic properties of LCNs with particular attention to their possible applications in the design and manufacture of integrated circuits.

Rensselaer Nanotechnology Center
Rensselaer Polytechnic Institute
110 8th Street
Troy, NY 12180-3590
Tel: (518) 276-6000
URL: http://www.rpi.edu/dept/nsec/
E-mail: rwsiegel@rpi.edu

The Rensselaer Nanotechnology Center was established in March 2001 to carry out research on advanced materials and coatings, biosciences and bionanotechnology, nanoelectronics, microelectronics, nanosystems, and the impact of nanotechnology on industry and society. Specific areas of concern involve more effective drug delivery systems, high capacity energy and information storage devices, and new flame-retardant materials for use in airplanes and automobiles. Later in 2001, RPI was selected by the National Science Foundation as one of its Nanoscale Science & Engineering Centers (Center for Directed Assembly of Nanostructures). The center is a collaborative effort that also includes as partners the University of Illinois at Urbana Champagne, Eastman Kodak, Philip Morris, IBM, ABB U.S., and Albany International.

Self-Assembly and Macromolecular Electronics Focus Group
Room 07, Kinard Hall
Clemson University
Clemson, SC 29634
Tel: (864) 656-3311
URL: http://virtual.clemson.edu/groups/nanotech/
E-mail: webmaster @hubcap.clemson.edu

The Self-Assembly and Macromolecular Electronics Focus Group studies the effects of large-scale self-assembly processes on the electronic structure of large molecular clusters. The special interest topics within this group are focused on the studies of (1) self-assembly in high-molecular-weight polymers, (2) micelle structures, and (3) carbon nanotubes. In the last-mentioned area, research is being conducted to determine how changes in nanotube structure affect electronic structures of the material.

Thermal Spray Lab
Department of Materials Science & Engineering
State University of New York at Stony Brook
Stony Brook, NY 11794-2275
Tel: (516) 632-8510
Fax: (516) 632-7878
URL: http://doll.eng.sunysb.edu/tsl/nano/nanozrpowdr.html

The Thermal Spray Lab uses a high-temperature jet spray to atomize and deposit nanosize particles on the surface of a material. By using this technique, researchers have been able to produce and manipulate ultrasmall particles that can be collected as a powder or deposited on a substrate.

Chapter Eleven

Print Resources

In many respects, molecular nanotechnology is a young field. One consequence of that fact is that the number of books, journals, and articles dealing specifically with the subject is relatively small. Many print resources refer incidentally to the subject, whether authors indicate that connection or not. In addition, print references to "top-down" nanotechnology are extensive. In fact, there are too many to include in this chapter, which focuses primarily (if not exclusively) on references to molecular nanotechnology.

The other problem created by the immaturity of nanotechnology as a science is that there is relatively little literature for the general reader. Still lacking many of the objects and tools from which a mature science of nanotechnology will grow, books and articles tend to repeat most of the basic predictions and issues outlined by K. Eric Drexler in his earliest works, such as *Engines of Creation.*

One should expect the number of print resources on nanotechnology to grow rapidly in the near future as the science itself evolves and matures. The selections presented here should be viewed primarily as a good general introduction to molecular nanotechnology. Readers who wish to become informed about later developments in the field should consider focusing also on Internet and other nonprint sources listed in Chapter 12.

Journals

Only a handful of journals dealing exclusively and specifically with the subject of nanotechnology exists. They are listed in the first section. Following that section are some journals of general interest and others dealing with specific fields of science. These journals often carry articles dealing with research in nanotechnology.

Nanotechnology Journals

Journal of Nanoparticle Research (JNP)
Kluwer Academic Publishers
PO Box 17
3300 AA Dordrecht, The Netherlands
and
101 Philip Drive
Norwell, MA 02061
Tel: (781) 871-6600
Fax: (781) 871-6528
URL: http://www.wkap.nl/journals/nano
E-mail: kluwer@wkap.com

JNP calls itself "the first interdisciplinary journal devoted to nanoparticle science and technology." It is available in print form and electronically online via Kluwer Online (http://www.wkap.nl/kaphtml.htm/HOMEPAGE).

Nano Letters
American Chemical Society Member & Subscriber Services
PO Box 3337
Columbus, OH 43210
Tel: (800) 333-9511
Fax: 614-447-3671
URL: http://pubs.acs.org/journals/nalefd/about.html
E-mail: service@acs.org

Nano Letters began publication in January 2001 as a monthly journal devoted to fundamental research in all branches of the theory and practice of nanoscience and nanotechnology. Specific areas of interest for the journal include the physical, chemical, and biological synthesis and processing of organic, inorganic, and hybrid nanoparticles; modeling and simulation of synthetic, assembly, and interaction processes; and realization and application of novel nanostructures and nanodevices.

Nanotechnology
Institute of Physics
Dirac House, Temple Back
Bristol, UK BS1 6BE
Also available from
American Institute of Physics
American Center for Physics
One Physics Ellipse
College Park, MD 20740-3844
Tel: 301-209-3100
Fax: 301-209-0843

Nanotechnology first was published in 1990 under the auspices of Great Britain's Institute of Physics. It now is jointly sponsored and published by the American Institute of Physics. The journal is devoted to the publishing of experimental and theoretical research or original synthesis or analysis of nanoscale objects. It includes research on "top-down" and "bottom-up" approaches to nanotechnology.

Many journal articles are probably comprehensible to upper-level high school students and should be read to appreciate the level of progress in the field. It is clear from the research reported that at least some mainstream researchers are taking Drexler's views seriously and framing their research within the context of a search for assemblers and other fundamental nanoscale devices. Some sample titles of articles in recent issues include "Constructing Nanomechanical Devices Powered by Biomolecular Motors," "Molecular Shuttles: Directed Motion of Microtubules along Nanoscale Kinesin Tracks," "Three-dimensional Manipulation of Carbon Nanotubes under a Scanning Tunneling Microscope," "Nanostressing and Mechanochemistry," and "Robust Self-assembly Using Highly Designable Structures."

Nanotechnology and Microengineering Progress
STICS, Inc.
9714 South Rice Avenue
Houston, TX 77096
Tel: (713) 723-3949

The emphasis of this journal tends to be on microengineering and "top-down" approaches to nanotechnology. Some of the topics covered are microchips and semiconductor technology, micromachines, genetic engineering and biotechnology, sensors, and microstructures and micromechanics. This journal is designed for professionals in the field.

Nanotechnology Industries Newsletter
#1518, 4739 University Way, NE
Seattle, WA 98105
URL: http://www.nanoindustries.com
E-mail: nanogirl@halcyon.com

The first issue of this newsletter appeared in June 1999. It is published by Gina "Nanogirl" Miller and is "directed towards the advancing science of nanotechnology." Some of the articles that have appeared in the first two issues include "Interview with Forrest Bishop," "The Problem of Explaining Nanotechnology to Others," and "HDL's for Nanomachinery." "Nanotechnology in the News" and "Nanotechnology Web Links" are regular features of the publication.

NanoTechnology Magazine
3 Degrees Kelvin Publishing, Inc.
4451 Sierra Drive
Honolulu, HI 96816
Tel: (808) 842-0914
Fax: (808) 842-0934
URL: http://nanozine.com
E-mail: info@nanozine.com

NanoTechnology Magazine is the first magazine to have molecular nanotechnology as its primary goal. It is published in print and electronic forms; the latter is available at its website, http://nanozine.com. The monthly publication provides general information about molecular nanotechnology, carries articles about important individuals in the field, surveys ongoing research, and outlines many of the issues likely to arise as the result of advances in molecular nanotechnology. The articles are well written and easily within the scope of any intelligent high school student. The web version of the magazine also carries many articles on topics related to molecular nanotechnology, such as a biography of K. Eric Drexler; a report on Zyvex, the first molecular nanotechnology company; graphics of nanoscale devices; and the ecological significance of molecular nanotechnology. The magazine also sponsors a NanoComputer Dream Team, an informal group of individuals whose goal is to build the first nanoscale computer.

Scanning
PO Box 832
Mahwah, NJ 07430-0832

Tel: (800) 443-0263

URL: http://www.scanning-fams.org

E-mail: scanning@fams.org

Scanning is an interdisciplinary journal intended for the rapid exchange of information among scientists interested in all phases of scanning electron, probe, and optical microscopy and applications of the technology.

Virtual Journal of Nanoscale Science & Technology (VJNST)

URL: http://ojps.aip.org/journals/doc/VIRT01-home/about.html.

VJNST is a weekly online journal that carries articles on nanoscale science and technology previously published in a variety of print journals, such as those of the American Physical Society, the American Institute of Physics, the American Association for the Advancement of Science, the Acoustical Society of America, and the American Vacuum Society. Articles are available at no charge to subscribers of a journal being accessed and can be downloaded by nonsubscribers for a moderate charge.

Other Journals

Journal of Computer-Aided Molecular Design

Kluwer Academic Publishers

Journals Department

PO Box 322

3300 AH Dordrecht, The Netherlands

Tel: (+31) 78 639 23 92

Fax: (+31) 78 654 64 74

URL: http://www.wkap.nl/

E-mail: services @wkap.nl

The early stages of research on molecular nanotechnology have focused to a significant extent on the design of molecular-scale devices with potential as assemblers, replicators, or other nanoscale devices. Elegant software programs have been developed that maximize the power of this tool. The *Journal of Computer-Aided Molecular Design* is written for professionals in the field, but it provides a fascinating overview of the kinds of work that can be done with this approach.

Journal of Molecular Modeling

Springer-Verlag New York, Inc.

PO Box 2485

Secaucus, NJ 07096

Fax: (201) 348-4505
URL: http://www.ccc.uni-eerlangen.de/jmolmod/index.html
E-mail: orders@springer-ny.com

This journal is intended for chemists whose primary interest is molecular modeling. It is available in print, Internet, and CD-ROM versions, the first journal in chemistry adapted to all three formats. In addition to published papers, the journal includes enhanced abstracts (short slide shows outlining the contents of a paper), communications with a turnaround time of less than 3 weeks, and reviews and e-mail discussions about topics in the field.

Journal of Vacuum Science & Technology
Subscription Fulfillment Department
SLACK, Inc.
6900 Grove Road
Thorofare, NJ 08086
Tel: (856) 848-1000
Fax: (856) 853-5991

The *Journal of Vacuum Science & Technology* is an official publication of the American Vacuum Society. It carries research papers, brief reports, comments, shop notes, review articles, and critical reviews. It is published 12 times a year in two sections, known as *JVST A* and *JVST B*. Both journals are designed for professionals in the field, although some general articles are written at a level that can be understood by high school students and the general public. The journals focus on research in the fields of vacuums, surfaces, films, microelectronics, and nanometer structure.

Molecular Engineering
Kluwer Academic Publishers
Journals Department
PO Box 322
3300 AH Dordrecht, The Netherlands
Tel: (+31) 78 639 23 92
Fax: (+31) 78 654 64 74
URL: http://www.wkap.nl/journalhome.htm/0925-5125
E-mail: services @wkap.nl

Molecular Engineering is an interdisciplinary journal intended for researchers working on the design, characterization, and application of molecules and molecular materials with specially designed biological, chemical, and physical properties. Some articles in recent volumes have dealt with the geometry of multi-tube carbon clusters and their electrical properties, tensile strength of thin films, and charge transfers in certain types of organic molecular structures.

NASA Tech Briefs
Associated Business Publications
317 Madison Avenue
New York, NY 10017-5391
Tel: (212) 490-3999
Fax: (212) 986-7864
URL: http://www.nasatech.com/
E-mail: Damiana@abptuf.org

NASA Tech Briefs is a system of communication developed by NASA to meet a U.S. Congress mandate that the agency report to industry all new, commercially significant technologies developed during NASA research and development. The journal first was published in the 1960s as single-sheet reports, then was published in its present form in the 1970s.

The journal is currently a monthly publication that contains articles written by engineers and scientists who have worked on new technologies in electronics, the physical sciences, materials science, computer software, mechanics, machinery and automation, manufacturing and fabrication, mathematics and information science, and life sciences. The journal also includes articles describing spinoffs of NASA developments, news briefs, and applications stories. The journal is of interest to those involved with molecular nanotechnology primarily because of the research being conducted in this field at Ames Research Center at Moffett Field in California.

Nature
Nature America, Inc.
345 Park Avenue South
10th Floor
New York, NY 10010-1707
Tel: 1-800-524-0384
Fax: 1-615-377-0525
URL: http://www.nature.com
E-mail (subscriptions): subscriptions@nature.com

Nature is the United Kingdom counterpart of *Science* in the United States, which is a highly respected journal covering all areas of science. It provides news and research reports that are accessible to the general reader with a moderate background in science along with more esoteric reports on research in all fields of science.

The journal now publishes specialized magazines in the fields of medicine (*Nature Medicine*), genetics (*Nature Genetics*), biology (*Nature Structural Biology*), and biotechnology (*Nature Biotechnology*). The family of *Nature* journals also is available on-line. This URL is open to the general public and provides weekly tables of contents, summaries, news articles, access to *Nature* archives, and job information. As is the case with *Science*, *Nature* is one of the most important journals for general readers interested in nanotechnology not only because of reports on original research it contains, but also because of the overview it provides on developments in the field and its relationship to other fields of science and technology and to industry, politics, and other fields of concern.

Precision Engineering—Journal of the International Societies for Precision Engineering and Nanotechnology
Elsevier Science
655 Avenue of the Americas
New York, NY 10010-5107
Tel: (212) 633-3730
Fax: (212) 633-3680
URL: http://www.elsevier.com/
E-mail: usinfo-f@elsevier.com

Precision Engineering is a joint publication of the American Society for Precision Engineering, the Japan Society for Precision Engineering, and the European Society for Precision Engineering and Nanotechnology. It publishes papers on high-accuracy engineering, metrology, and manufacturing at the whole range of scales, from macroscopic to atom-based nanotechnology.

Protein Engineering
Oxford University Press
Journals Subscription Department
Great Clarendon Street
Oxford OX2 6DP
UK
Tel: +44 (0) 1865 267907
Fax: +44 (0) 1865 267485
E-mail: jnl.info@oup.co.uk
or, in the United States:
Tel: (919) 677-0977; (800) 852-7323
E-mail: jnlorders@oup-usa.org

One of the important "enabling technologies" in the development of molecular nanotechnology is protein engineering. An *enabling technology* is a field in which tools and products are likely to have direct relevance to the development of molecular nanotechnology itself. In this instance, researchers long have recognized that naturally occurring proteins can serve as models for the kind of assemblers that will be needed in molecular nanotechnology. A great deal of research is going on to understand the mechanisms by which the various levels of protein structure, especially the folding of proteins, develop. The knowledge learned from this research should be applicable not only to the invention of artificial and modified natural proteins, but also to nonorganic analogues of natural proteins. The journal is intended for specialists in the field.

Science
American Association for the Advancement of Science
1200 New York Avenue, N.W.
Washington, D.C. 20005
Tel: 1-202-326-6400
URL: http://www.aaas.org
E-mail: webmaster@aaas.org

Science is one of the two premier scientific journals for general readers around the world (*Nature* being the other). The magazine contains seven major sections: News and Comment, Research News, Perspectives, Articles, Reports, Technical Comments, and a Web Feature. Its regular departments include a brief overview of the magazine, "This Week in *Science*"; an Editorial; Letters to the Editor; brief news notes, "Random Samples"; Book Reviews; and information on new technology and products, "Tech.Sight: Products."

Science has an extensive Internet website that includes two major sections, Science NOW and Science's Next Wave. Science NOW includes Current Stories and Week in Review stories from the world of science, along with a search function that allows a limited search of the journal's archives and current issues. Next Wave includes sections such as "Going Public," an open forum for discussion of topics in science; "New Niches," a feature on alternative careers in science; "In the Loop," the page's news section; "Tooling Up," a column on science career advice; "Signposts," links to other web pages; and "Wavelengths," an opportunity to interact with editors of *Science*.

Science is of particular importance to those interested in nanotechnology not only because it carries reports of basic research in the area, but also because the journal provides a guide as to overall progress in the field and indications of its importance to industrial, governmental, and political leaders.

Single Molecules
Wiley–VCH Reader Service
PO Box 10 11 61
D-69451 Weinheim
Germany
Tel: +49 (0) 6201 / 606 147
Fax: +49 (0) 6201 / 606 117
URL: http://www.wiley-vch.de/berlin/journals/singmol/index.html
E-mail: subservice@wiley-vch.de

The first issue of *Single Molecules* appeared in January 2000. The journal is designed to "provide researchers with a broad overview of current methods and techniques, recent applications and shortcomings of present techniques in the field of single molecules." Some of the topics to be included in the journal are single atoms and nanoparticles, ultrasensitive detection techniques, and methods of manipulating and modifying single particles.

Small Times
755 Phoenix Drive
Ann Arbor, MI 48108
Tel: (734) 994-1106
Fax: (734) 994-1554
URL: http://www.smalltimes.com
E-mail: info@smalltimes.com

Small Times was created in September 2001 with a focus on developments in the fields of micro- and nanoscale science and technology. It carries stories not only on research and development in the field, but also on business opportunities created by these developments. It includes articles on recent business developments related to nanoscience and nanotechnology, patent listings, job opportunities, a calendar of forthcoming events, and a message board. Its website is an unusually valuable source of information on these and related topics.

Supramolecular Chemistry
Journals Fulfillment Department
Butterworth-Heinemann
313 Washington Street, 4th Floor
Newton, MA 02158

Supramolecular chemistry is a relatively new area of research dealing with the chemistry of particles consisting of clusters of a relatively small number of atoms or molecules, such as quantum dots. Many of the articles in this journal have relevance to the field of molecular nanotechnology, although they are written primarily for researchers in the field.

Books

Bard, Allen J. *Integrated Chemical Systems: A Chemical Approach to Nanotechnology.* New York: Wiley & Sons, 1994. ISBN 0-471-00733-1.

Integrated Chemical Systems is based on the Baker Lectures given in 1987 at Cornell University. The author includes some interesting speculations about the basic principles of molecular nanotechnology. He defines an integrated chemical system as a system with many different parts "designed and arranged for specific functions to carry out specific reactions or processes." He begins with a description of biological systems that fit this definition, then goes on to show how human-made systems with properties similar to those of biological systems can be designed. For the general reader, Chapter 2, "Construction of Integrated Chemical Systems," may be of greatest interest.

Bimberg, D., M. Grundmann, and N. N. Ledentsov. *Quantum Dot Heterostructures.* Chichester, UK: John Wiley & Sons, 1998. ISBN 0-471-97388-2.

Quantum dots are nanometer-size semiconductor structures, currently one of the most active fields of nanoscale research. This book provides a general introduction to the subject and a review of current research for advanced undergraduates, graduates, and professionals in the field. The major topics covered are the history of quantum dot development, fabrication techniques, self-organization processes, growth and structural characteristics of self-organized quantum dots, electronic and optical properties of quantum dots, and photonic devices.

Chow, Gan-Moog, and Kenneth E. Gonsalves, eds. *Nanotechnology: Molecularly Designed Materials.* Washington, DC: American Chemical Society, 1996. ISBN 0-841-23392-6.

In contrast to the sense that may be suggested by the title of this volume, it has nothing to do directly with molecular nanotechnology. It deals, instead, with more traditional topics, such as the properties of vapor phase, metal colloids in polymers, semiconductors, nanocomposites, ceramics, and sol-gels. The book contains a collection of papers offered at a symposium sponsored by the Division of Polymeric Materials, Science and Engineering, Inc., at the 210th National Meeting of the American Chemical Society, August 20–24, 1995.

Crandall, BC. *Nanotechnology: Molecular Speculations on Global Abundance.* Cambridge, MA: MIT Press, 1996. ISBN 0-262-03237-6.

Crandall provides a general introduction to the subject of nanotechnology in the first chapter of this book. That introduction is followed by three major sections, on the applications of nanotechnology within the human body, outside the human body, and in some highly technical situations. Some of the applications considered include cosmetic nanosurgery, tooth replacement, personal computers, and hobbies. The book is written at a level that makes it accessible to any intelligent layperson and provides sound scientific and technological information along with some interesting prognostications about the potential uses of nanotechnology. A useful collection of notes for each chapter adds to value of the book.

Crandall, BC, and James Lewis, eds. *Nanotechnology: Research and Perspectives.* Cambridge, MA: MIT Press, 1992. ISBN 0-262-03195-7.

This volume is a collection of papers presented at the First Foresight Conference on Nanotechnology, held in Palo Alto, California, in October 1989. Papers are divided into three major sections: Molecular Systems Engineering, Related Technologies, and Perspectives. Examples of topics included in the first section are "Atomic Imaging and Positioning," "Design and Characterization of 4-Helix Bundle Proteins," and "Strategies for Molecular Systems Engineering."

The Related Technologies section of the book includes papers on molecular electronics, quantum transistors, and fundamental problems to be solved in the development of molecular systems. The final section of the book is devoted to possible applications of nanotechnology along with social, ethical, and other issues that may arise as the result of developments in this field. Included among the papers in this section are discussions of medical applications of nanotechnology, economic consequences of the field, risks and dangers posed by nanotechnology, and methods for dealing with issues raised by research in molecular nanotechnology.

Two useful appendices provide a reprint of Eric Drexler's article on "Machines of Inner Space" from the Encyclopedia Britannica's *1990*

Yearbook of Science and the Future and of Richard Feynman's 1959 lecture, "There's Plenty of Room at the Bottom: An Invitation to Enter a New Field of Physics."

Darling, David. *Beyond 2000: Micromachines and Nanotechnology: The Amazing New World of the Ultrasmall.* Morristown, NJ: Silver Burdett Press, 1995. ISBN 0-3822-4953-4.

A work intended for readers in grades 5 to 9, this book outlines the history of the increasing miniaturization of material design and construction. The emphasis seems to be on microtechnology, although some introduction to the subject of molecular nano-technology also is provided.

Day, David E., et al. *Cyberlife!* Indianapolis: Sams Publishing, 1994. ISBN 0-672-30491-0.

Cyberlife! is a general-interest book written at the high school level and aimed at the general reader. Chapter 23 of the book outlines most of the basic arguments about nanotechnology. This book is a good introduction to the subject, although limited in scope and now outdated.

Drexler, K. Eric. *Engines of Creation.* New York: Anchor Books, 1986. ISBN 0-385-19973-2 (pbk.).

Engines of Creation is essential reading for anyone interested in nanotechnology. Its author probably should be considered the Father of Nanotechnology, and this work is his first effort to present and explain the subject to a general audience.

The book is an interesting mixture of nanotechnology and a host of other subjects on which the field will have impact and which, in turn, will determine the nature and direction of research in nanotech-nology. Drexler lays out the fundamental concepts of nanotechnol-ogy in Chapters 1 and 4, in which he describes and explains such basic ideas as assemblers and disassemblers, replicators, and nano-computers. He also shows in detail how nanomachines already exist in nature in the form of protein and nucleic acid molecules and how the existence of these molecules provides models for future nano-machines. He then describes "second-generation nanotechnology," in which human researchers will build structures similar to but far more complex and flexible than biological molecules. In Chapter 4, Drexler outlines the methods by which nanomachines will be able to construct a virtually limitless number and variety of useful machines, from pins and needles to tanks and skyscrapers.

Other chapters deal with specific and important applications of nanotechnology. Chapter 6 explains how nanotechnology will be able to revolutionize human exploration of and research in space. Chapter 7 discusses the application of nanotechnology to human health and longevity, forecasting a future in which disease, poor health, and injuries are either eliminated or made easily curable.

Drexler's book is important not only for its explication of nanotechnology, but also for its analysis of a host of social, political, economic, ethical, and psychological problems associated with science and technology. His work in nanotechnology from the outset has convinced him that this new field of research has enormous potential for good and evil and that humans will have to think about and plan for the new revolution in nanotechnology long before products begin appearing in the real world.

As a result, Drexler talks in Chapter 3 about some general principles that apply in the prediction and projection of future developments in any field of science and technology. He also presents some specific recommendations for the ways in which future developments in nanotechnology can be controlled to ensure that they do not result in catastrophes beyond anything humans have ever seen. He suggests that these systems of control will require technological and social fixes of types that currently are not available but that must be developed eventually if the hopes and promises of nanotechnology are to be met.

Engines of Creation is now available in full on the Internet at http://www.foresight.org/ECO/. In 2000, the Foresight Institute announced the *Engines of Creation 2000* Project, a program aimed at updating Drexler's original book. The project will include the development of software that will allow structured discussion of the book on the Internet. Additional information about the *Engines of Creation 2000* Project is available at http://www.foresight.org/engines/index.html.

Drexler, K. Eric. *Nanosystems: Molecular Machinery, Manufacturing, and Computation.* New York: John Wiley & Sons, 1992. ISBN 0-471-57547-X.

Nanosystems was Eric Drexler's third major book (after *Engines of Creation* and *Unbounding the Future*) and the first major technical exposition of molecular nanotechnology. In this work, he discusses many fundamental chemical and physical issues involved in the development of nanosize devices. In Part I of the book, he deals with issues such as classic physical systems and scaling laws as they apply to the nanoscale level, quantum and kinetic effects at the nanolevel, molecular dynamics, positional uncertainty, errors in and damage to nanoscale devices, heat production and dissipation, and mechanosynthesis.

Part II focuses on specific components and systems encountered at the nanometer scale. Drexler presents many models of familiar devices, such as gears, clutches, ratchets, drives, motors, generators, and pumps, as they would look if they consisted of only a few dozen or few hundreds of atoms or molecules. Part III is devoted to a discussion of some implementation strategies that might be used in the development of molecular nanotechnology, strategies such as biotechnology, solution synthesis, and mechanosynthesis.

Two important appendices are devoted to a discussion of the methodological issues involved in theoretical applied science (such as that on which the early stages of molecular nanotechnology is based) and research from related fields, such as biotechnology and microtechnology, chemistry, protein engineering, and molecular biology.

The book is important in the early history of molecular nanotechnology because the author addresses many of the scientific and technical questions raised explicitly or implicitly by many of his critics. Some authors have suggested that it marks the turning point in the development of molecular nanotechnology, providing a technical basis for what previously had been regarded widely as a science fiction endeavor. *Nanosystems* was awarded the 1992 Association of American Publishers award for the year's best computer book.

Drexler, K. Eric, and Chris Peterson, with Gayle Pergamit. *Unbounding the Future: The Nanotechnology Revolution.* New York: William Morrow, 1991. ISBN 0-688-09124-5.

Unbounding the Future is a popular version of *Engines of Creation* written in nontechnical terms for the general reader. It consists of short sections with special attention to the potential and the potential problems posed by the rise of molecular nanotechnology.

DuCharme, Wesley M. *Becoming Immortal: Nanotechnology, You, and the Demise of Death.* Evergreen, CO: Blue Creek Ventures, 1995. ISBN 0-9646-2820-1.

One of the most intriguing potential results of a fully developed molecular nanotechnology is the total elimination of disease and other forms of illness and, hence, an indefinite extension of life. Because of this potential, many individuals interested in molecular nanotechnology have explored the question of what eternal life might mean for humans. This book provides an interesting general introduction to that question.

Endo, M., M. S. Dresslhaus, and S. Iijima, eds. *Carbon Nanotubes.* New York: Elsevier Science, 1996. ISBN 0-080-42682-4.

Arguably the most important book on carbon nanotubes to have been written during the early 1990s, this work covers most of the fundamental information about these interesting structures known at the time. The chapters cover basic topics, such as the production of carbon nanotubes from vaporized carbon; electrical effects on nanotube growth physics of carbon nanotubes; carbon nanotubes with single-layer walls; scanning tunneling microscopy of carbon nanotubes; properties of buckyballs and derivatives; electronic, mechanical, and thermal properties of carbon nanotubes; and nanoparticles and filled nanocapsules.

Freitas, Robert A., Jr. *Nanomedicine.* Austin, TX: Landes Bioscience, 1999. ISBN 1-5705-9645-X.

Volume 1 of this projected three-volume work deals with the topic of "Basic Capabilities." The author explains that his goal is to

> describe the preliminary technical issues involved in medical nanodevice design, along with some details of the required foundational technical competencies. These competencies include the ability to sort and transport molecules; chemical, acoustic, and other physical sensors; shape, texture, and compositional control of external surfaces; in vivo power storage, generation, and transmission; communication among nanorobots, and between doctors, patients, and in vivo nanorobots; navigation throughout the body and inside individual cells; manipulation and locomotion at the microscopic level; nanocomputation, timers and nanoclocks; cryothermal operations; and defensive cellular armaments.

Volumes 2 ("Systems and Operations") and 3 ("Applications") are planned for completion "over the course of several years." Volume 1 is currently available online at http://www.nanomedicine.com. Other volumes also will be made available on the same site as they are completed.

Fujimasa, Iwao. *Micromachines: A New Era in Mechanical Engineering.* Oxford: Oxford University Press, 1996. ISBN 0-198-56513-5.

This book consists of two sections; the first and by far the longer deals with micromachines. As their name suggests, such machines are typically a few microns in size. The second section deals with nanomachines that can be built from the top down using micromachines. The author is familiar with Drexler's work and uses extensive analogies from biological systems in discussing microscale and nanoscale machines.

Halperin, James L. *The First Immortal.* New York: The Ballantine Publishing Group, 1998. ISBN 0-345-42092-6.

Halperin has written a science fiction novel that incorporates many of the fundamental ideas of molecular technology. A review in the newsletter *Foresight Update* (volume 33) observes that the story is "a thoroughly engaging story of a plausible future for humanity. It is firmly anchored in our current knowledge of science—there is no fantasy, no need to postulate some *ad hoc* discovery of new physics or some far-fetched technology with no basis in current science."

In Halperin's story, Benjamin Franklin Smith, a physician, dies of a heart attack in his early sixties. He has previously made arrangements to have his body frozen after his death, placing him in suspended animation until some unspecified future date when medical science will have discovered ways to revive him and restore his life to him.

A significant portion of the book is devoted to the problems faced by Smith's surviving relatives when they learn of his plans and of the new challenges faced by Smith and other relatives when they are reunited eight decades after his natural "death." The book is an intriguing story that places into a real-life context many of the social, emotional, and psychological issues raised by the development of molecular nanotechnology.

Hingle, H. T., and J. W. Gardener, eds. *From Instrumentation to Nanotechnology.* New York: Gordon & Breach, 1999. ISBN 2-8812-4794-6.

This volume consists of 16 lectures presented at an advanced vacation school on instrumentation and nanotechnology held in Warwick, England, in September 1990. The focus of the presentations is on the design, manufacture, and assessment of components with dimensions in the 0.1- to 100-nm range. Some of the chapters included are "Ultrasonic Sensors," "Use of Energy Beams for Ultrahigh Precision Processing of Materials," "Nanotechnology," "Nanoactuators for Controlled Displacements," and "High Precision Surface Profilometry: From Stylus to STM."

Hoch, Harvey C., Lynn W. Jelinski, and Harold G. Craighead, eds. *Nanofabrication and Biosystems: Integrating Materials Science, Engineering, and Biology*. Cambridge: Cambridge University Press, 1996. ISBN 0-521-46264-9.

The articles in this collection report on new approaches to microfabrication and nanofabrication with special applications to biological materials. Some of the topics included are lithographic techniques, materials etching, microsensors and microactuators, self-assembly of monolayers, cellular engineering, and microcontrol of neural growth.

Interrante, Leonard V., and Mark J. Hampden-Smith, eds. *Chemistry of Advanced Materials: An Overview*. New York: Wiley-VCH, 1998. ISBN 0-471-18590-6.

This book is a collection of articles that appeared originally in the August 1996 issue of the journal *Chemistry of Materials*. The two chapters of special interest in the area of molecular nanotechnology are Chapters 7 ("Nanoparticles and Nanostructured Materials") and 8 ("Nanoporous Materials"). Both chapters focus on the properties and behavior of materials a few nanometers in size, although there is little specific mention of molecular nanotechnology.

Koruga, Djuro. *Fullerene C60: History, Physics, Nanobiology, Nanotechnology*. Amsterdam: North-Holland, 1993. ISBN 0-444-89833-6.

This book is fascinating and unorthodox in that it combines a sophisticated analysis of the chemistry and physics of fullerenes with some philosophical speculations (such as the special functions of spatial and temporal icosahedrons) that predate modern science by many centuries. Many parts of the book are readable by high school students.

Krummenacker, Markus, and James Lewis, eds. *Prospects in Nanotechnology: Toward Molecular Manufacturing*. New York: John Wiley & Sons, Inc., 1995. ISBN 0-471-30914-1.

Prospects in Nanotechnology is required reading for anyone interested in gaining a broad general introduction to the status of molecular nanotechnology in the early 1990s. The book contains the proceedings of the First General Conference on Nanotechnology held in Palo Alto, California, November 11–14, 1992. Most of the major pioneers in the field were present as presenters or participants.

The opening chapter, by K. Eric Drexler, provides background information on and an introduction to the general topic of nanotechnology.

The next few chapters deal with technical issues in the design and production of nanotechnology devices, written at a level accessible to most general readers. These chapters emphasize challenges and opportunities in biotechnology and molecular manufacturing. Michael Pinneo describes in his chapter methods for growing the diamonoid (diamond-like) materials that will be needed for nanostructures. Michael Pique discusses methods that have been developed for the production of biological molecules with functions needed in nanotechnology.

In some regards, the most helpful chapter in the book for the general reader is "Virtual Molecular Reality," by Marvin Minsky. In this chapter, Minsky presents the fundamental concepts of molecular nanotechnology and all of its promises, while dealing clearly with the objections that have been raised about the field. Later chapters deal with many ancillary issues, such as the politics of nanotechnology in the United States, the role of venture capitalism in the development of nanotechnology, nanotechnology in Japan, and possible roles for nanotechnology in space science.

Kuhn, Thomas S. *The Structure of Scientific Revolutions*, 2nd edition. Chicago: University of Chicago Press, 1986. ISBN 0-226-45807-5.

Kuhn's book is one of the classics in the history and philosophy of science. In it, he explores the circumstances under which major paradigm shifts take place in science. A *paradigm* is a general framework within which the research in any particular field of science is conducted. An example of a major paradigm shift in the history of science was the Copernican Revolution in astronomy, which forced astronomers to stop thinking of the Earth as the center of the Universe.

The Structure of Scientific Revolutions is important reading for those interested in molecular nanotechnology because it helps explain the philosophical controversy that surrounds research in the field. Many highly respected scientists hold little or no hope that molecular nanotechnology will ever achieve the goals outlined by Eric Drexler and his supporters. They say that the laws of science argue against the construction of assemblers, replicators, and other such nanoscale devices.

Arguments of this kind *always* have been offered against major breakthroughs in science, however (such as the Copernican Revolution). In such cases, naysayers often are left behind as the new revolution in science sweeps away the old paradigm in which they are working to make way for the new paradigm. What is most interesting about such patterns is that the same naysayers often make major and critical contributions to the framework of the new paradigm.

Of course, every new idea in science does not lead to a new paradigm. Many ideas are erroneous and, over time, are proved to be of little or no value to the advancement of scientific knowledge. The "discovery" of cold fusion in 1989 is an example of such an occasion. Many researchers thought that a whole new field of nuclear physics was opening up with this "discovery." It seems in retrospect, however, that such hopes were overly optimistic, if not simply wrong.

One cannot make a judgment about the value of research along new lines until a period of time has passed. In the case of molecular nanotechnology, one still cannot say what successes it may have in the future. It could lead to the greatest breakthroughs in the history of the physical sciences, or it could be a long path leading to a dead end. Anyone who is following the development of this story will want to be familiar with Kuhn's research on scientific revolutions to have a better perspective from which to judge current developments.

Kuzmany, Hans, Jörg Fink, Michael Mehring, and Siegmar Roth. *Molecular Nanostructures. Proceedings of the International Winterschool on Electronic Properties of Novel Materials.* Singapore: World Scientific 1998. ISBN 981-02-3261-6.

This collection of articles covers the electronic, optical, and mechanical properties of nanoscale structures. It is appropriate for the advanced college level.

Lampton, Christopher. *Nanotechnology Playhouse.* Corte Madera, CA: Waite Group Press, 1993. ISBN 1-878-73933-6.

This book is a simple introduction to nanotechnology with many illustrations to help explain and describe the written text. The book is accompanied by an IBM-compatible disk containing "multimedia nanomachine simulations."

Lee, Stephen C., and Lynn M. Savage, eds. *Biological Molecules in Nanotechnology: The Convergence of Biotechnology, Polymer Chemistry, and Materials Science.* Southborough, MA: IBC Libraries, 1998. ISBN 1-57936-109-9.

This book includes papers presented at a conference sponsored by International Business Communications, "Biological Approaches and Novel Applications for Molecular Nanotechnology," held in La Jolla, California, on December 8 and 9, 1997. The basic premise underlying the conference and the book, as defined by the editors, is that, first, "technology for building nanomachines with functional biological components is here today," and second, "no 'fatal flaw' has yet been identified for the nanobiological approach to nanotechnology." The book is divided into two major sections, the first dealing with enabling technologies for nanobiological devices and the second reviewing current applications of biologic nanotechnology.

Lyshevski, Sergey Edward. *Nano- and Microelectromechanical Systems: Fundamentals of Nano- and Microengineering.* Boca Raton, FL: CRC Press, 2001. ISBN 0-8493-0916-6.

This text provides a highly mathematical and rigorous review of basic principles involved in research and development in the fields of MEM and nanotechnology. The book is suitable for advanced students and professionals in the field.

Meyer, E., R. M. Overney, K. Dransfeld, and T. Gyalog. *Nanoscience: Friction and Rheology on the Nanometer Scale.* London: World Scientific Publishing Company, 1997. ISBN 9-8102-2562-8.

A technical volume for advanced undergraduate and graduate students, this book is an introduction to the use of friction force microscopy, a technique for examining the nature of surfaces with a modified scanning tunneling microscope. Chapters deal with topics such as "Normal Forces at the Atomic Scale," "Dissipation Mechanisms," "Nanorheology and Nanoconfinement," and "Friction Force Microscopy Experiments."

Nalwa, Hari Singh, ed. *Handbook of Nanostructured Materials and Nanotechnology.* San Diego: Academic Press, 2000. ISBN 0-125-13760-5 (set).

Handbook is a five-volume set of reference data on the synthesis and processing, spectroscopy and theory, and electrical and optical properties of quantum dots, nanotubes, nanorods, nanowires, nanofilms, self-assemblers, thin films, and other nanoscale structures and materials. The work is suitable for researchers in the field of "bottom-up" and "top-down" nanotechnology.

Nanotechnology Opportunity Report. Madrid: CMP Científica, 2002.

Nanotechnology Opportunity Report is a joint venture of CMP Científica; NRW, a public relations agency; and nABACUS, an Asian consulting and investment company. It provides a thorough analysis of opportunities in the growing nanotechnology market and in enabling and related technologies. Various chapters deal with topics such as "The Coming Nanotechnology Revolution," "Tools," "Analysis of Key Technologies and Players," "Industries Impacted by Nanotechnology," "Corporate Profiles," and "Research Center Profiles."

The book was scheduled to be released in February 2002 and can be ordered from CMP Cientifica (fax: +34 91 640 71 86 in Europe; by telephone: 650-827-7045 in the United States; or by e-mail at NOR@cmp-cientifica.com).

Naval Research Laboratory. *Nanoscience and Nanotechnology in Europe*. Washington, DC: Naval Research Laboratory. Publication NRL/FR/1003-94-9755.

This publication reviews work on "bottom-up" and "top-down" research in nanotechnology in Europe. It includes more than 350 references to specific projects that have been or are being conducted, primarily in France, Germany, Switzerland, and The Netherlands.

Nelson, M., and C. Shipbaugh. *The Potential of Nanotechnology for Molecular Manufacturing*. Santa Monica, CA: The RAND Corporation, 1995. ISBN 0-833-02287-3.

The RAND Corporation is a research institute that studies and issues reports on a wide variety of scientific, technological, social, and economic issues. In this book, the authors explore the possibilities of using nanotechnologies to construct machines of nanometer dimensions. They point out that it is too early in the development of this field to lay out a road map for progress in the field, although it should be possible to develop feasibility studies that would stimulate "near-term interim achievements of great merit." Such a strategy, the authors suggest, could prevent "technological surprises from foreign players."

The book is divided into five chapters: "Introduction," "Trends and Goals," "Developing Incremental Checkpoints," "National and International Research Efforts," and "Conclusions and Recommendations." An appendix lists research centers in nanotechnology and related areas by nation.

Nicolini, Claudio, and Sergei Vakula. *Molecular Manufacturing*. Dordrecht, The Netherlands: Kluwer Academic/Plenum Publishers, 1996. ISBN 0-306-45284-7.

This book includes papers presented at the 1994 International Workshops on Electronics and Biotechnology Advanced, held on the Isle of Elba, Italy, March 10–12 and September 5–15, 1994.

Regis, Ed. *Nano: The Emerging Science of Nanotechnology.* Boston: Little, Brown & Company, 1995. ISBN 0-316-73858-1.

The layperson can hardly do better than begin his or her study of nanotechnology with this book. Regis writes clearly and with an obvious passion for his subject. He weaves the strands of a biography of K. Eric Drexler with the primary elements in the rise and early history of nanotechnology in an enchanting manner.

Part I of the book introduces the subject of nanotechnology and outlines some of the early objections to the field (quantum mechanical considerations, the uncertainty principle, and thermal motion). He then discusses Richard Feynman's early paper on "There's Plenty of Room at the Bottom" and shows how Drexler originally became interested in the subject.

Part II outlines the development of molecular nanotechnology in Drexler's thinking and the cooperation and opposition he experienced in talking about the subject. Regis nicely suggests the possible "paradigm-shift" that may be taking place in science as a result of Drexler's ideas and describes the reaction of scientists in many fields to the new field of endeavor. The book includes a limited number of useful notes that include most of the fundamental references in the field as of the book's publication date.

Roco, Mihail C., and William Sims Brainbridge, eds. *Societal Implications of Nanoscience and Nanotechnology.* Dordrecht, The Netherlands: Kluwer Academic Publishers, 2001. ISBN 0-7923-7178-X.

Developments in nanoscience and nanotechnology promise to have some important effects on all aspects of human life. This book reports on a conference held in Arlington, Virginia, in March 2001 to discuss some of those implications. The book is divided into two major sections, the first of which includes chapters on "Nanotechnology Goals," "Nanotechnology and Societal Interactions," and "Social Science Approaches for Assessing Nanotechnology's Implications."

The second half of the book provides about three dozen statements on various aspects of the societal implications of nanoscale research on topics such as economic and political implications, science education; medical, environmental, space and national security issues; and social, ethical, legal, and cultural implications.

Shalaev, Vladimir M., and Martin Moskovits, eds. *Nanostructured Materials: Clusters, Composites, and Thin Films.* ACS Symposium

Series #679. Washington, DC: American Chemical Society, 1997. ISBN 0-841-23536-8.

This highly technical book is a collection of articles dealing with the physics and chemistry of nanoscale objects and materials and devices derived from them. The authors report on research from physics, chemistry, biology, and other areas of science.

Tanaka, K., T. Yamabe, and K. Fukui. *The Science and Technology of Carbon Nanotubes.* New York: Elsevier, 1999. ISBN 0-08-042696-4.

As the title suggests, this book provides a general overview of the latest information available on carbon nanotubes, including the purification and characteristics of single-walled and multiwalled carbon nanotubes and their physical properties, and the direction of research in the field.

Taniguchi, Norio, ed. *Nanotechnology: Integrated Processing Systems for Ultra-precision and Ultra-fine Products.* Oxford: Oxford University Press, 1996. ISBN 0-198-56283-7.

This book is a collection of technical articles dealing with the "top-down" approach to nanotechnology. Suitable for advanced college level.

Teich, Albert H., et al. *AAAS Science and Technology Policy Yearbook 2001.* Washington, DC: American Association for the Advancement of Science, 2001. AAAS Publication Number 01–03S.

Part 3 of this book considers "Technology's Impact on Society," dealing both directly and indirectly with advances in nanoscale research. It includes Bill Joy's now-famous article, "Why the Future Doesn't Need Us," along with a reply by John Seely Brown and Paul Duguid ("A Response to Bill Joy and the Doom-and-Gloom Technofuturists") and an article by Michael M. Crow and Daniel Sarewitz on "Nanotechnology and Society Transformation."

Timp, Gregory, ed. *Nanotechnology.* New York: Springer Verlag, 1999. ISBN 0-387-98334-1.

Twenty-two contributing authors from atomic physics, microelectronics, polymer chemistry, and other fields outline current research on objects and processes at the submicron level. The articles also outline some potential developments in the field of nano-technology in which devices are manufactured with nanometer precision and single atoms are used for computation.

Von Neumann, John. *Theory of Self-Reproducing Automata*. Edited and completed by Arthur W. Burks. Urbana: University of Illinois Press, 1966. LCCN 63007246.

von Neumann explores the mathematical considerations involved in the design of machines that are capable of manufacturing other machines with specific design specifications, including machines that are exact copies of themselves. In this regard, von Neumann has written about devices that are the macroscopic equivalent of Drexlerian assemblers and replicators.

In von Neumann's theory, self-reproducing automata consist of two parts: a Universal Computer and a Universal Constructor. The Universal Computer contains a program that directs the operation of the Universal Constructor. The Universal Constructor has the capabilities of producing specific machines, including other copies of itself and additional copies of the Universal Computer. When the two original machines were created and the copying process initiated, unlimited numbers of both devices could be produced.

Articles

At this point in history, molecular nanotechnology has been mentioned in hundreds of articles in the popular press. Those mentioned here have been selected because they contain more recent or more comprehensive discussions of the subject. For a more exhaustive list of articles mentioning molecular nanotechnology, see the Foresight Institute website at http://www.foresight.org/Updates/Publications.html.

Aeppel, Timothy, "Think Small," *The Wall Street Journal*, 1 January 2000. Also on the Internet at http://interactive.wsj.com/ millennium/articles/SB944517534664949434.htm.

Aeppel provides a broad, general introduction to the subject of molecular nanotechnology and its potential impact on human society, including the business world. The article is especially interesting in that it made the case for a revolutionary form of technology for which there are still many skeptics in one of the most conservative business publications in the United States.

Becker, R. S., J. A. Golovchenko, and B. S. Swartzentruber, "Atomic-Scale Surface Modification Using a Tunnelling Microscope," *Nature*, 29 January 1987, 419–21.

The authors claim to have used the scanning tunneling microscope to bring about "the smallest spatially controlled, purposeful transformation yet impressed on matter" by moving individual silicon atoms around on the surface of a perfect silicon crystal. The method described in the article was later to become a standard procedure for manipulating atoms and clumps of atoms.

"Beyond Silicon: The Future of Computing," *Technology Review*, May/June 2000.

This special issue of the magazine focuses attention on the future of nanotechnology, with special reference to molecular electronics, quantum computing, computing with biological cells, and computing with DNA molecules.

Binnig, Gerd, and Heinrich Rohrer, "The Scanning Tunneling Microscope," *Scientific American*, August 1985, 50–56.

The inventors of the scanning tunneling microscope describe their invention and suggest some possible uses for it in the areas of physics, chemistry, and biology in this general-interest article.

Dagani, Ron, "NASA Goes Nano," *Chemical and Engineering News*, 28 February 2000, 36–38.

This cover story discusses the 5-day conference, "NanoSpace 2000," held in Houston in January 2000. The author notes that "the soul and passion of the meeting seemed to be focused squarely on the nano realm, the new frontier that promises revolutionary possibilities." The enthusiasm for nanotechnology among space scientists was particularly interesting because interest in the subject until the last few years has been minimal among most members of the space community. Some of the potential applications for nanotechnology in the space program mentioned at the conference include improved materials, instruments, and sensors; the development of robotic probes; and advanced computing systems.

Drexler, K. Eric, "Molecular Engineering: An Approach to the Development of General Capabilities for Molecular Manipulation," *Proceedings of the National Academy of Sciences*, September 1981, 5275–5278.

This article often is regarded as the first formal scientific presentation of the principles of molecular nanotechnology. It was written by Drexler shortly after he learned about Richard Feynman's talk on "There's Plenty of Room at the Bottom." Drexler recognized that it was essential for him to establish scientific priority for many of the ideas about molecular nanotechnology about which he had been thinking for some years. In the article, Drexler states that "the ability to design protein molecules will open a path to the fabrication of devices to complex molecular specifications." That capability, Drexler then points out, makes it possible to "outline a path to [a new goal in technology], a general molecular engineering technology." He then lays out most of the fundamental principles of nanotechnology on which he later elaborates in *Engines of Creation* and *Nanosystems* (see next entry).

Eigler, D. M., C. P. Lutz, and W. E. Rudge, "An Atomic Switch Realized with the Scanning Tunnelling Microscope," *Nature*, 15 August 1991, 600–603.

The authors of this article claim that they have invented the "prototype of a new class of potentially very small electronic devices which we will call atom switches." The switches work by using a scanning tunneling microscope to move a single xenon atom between the tip of the microscope and a nickel surface.

Eigler, D. M., and E. K. Schweizer, "Positioning Single Atoms with a Scanning Tunnelling Microscope," *Nature*, 5 April 1990, 524–26.

This article contains a description of the now-famous experiment in which a scanning tunneling microscope is used to position 35 xenon atoms on the surface of a nickel crystal to spell out the IBM logo.

Freitas, Robert A., Jr., "Say 'Ah!'," *The Sciences*, July/August 2000, 26–31. Also on the Internet at http://www.nyas.org/membersonly/sciences/sci0007/freitas_body.html

The "father of nanomedicine," Robert Freitas, presents an easily understood and comprehensive review of the potential for developments in medicine made possible by molecular nano-technology. For someone who has read nothing else on nano-medicine, this is as good an introduction as can be found.

Garfinkel, Simson, and K. Eric Drexler, "Critique of Nanotechnology: A Debate in Four Parts," *Whole Earth Review*, Summer 1990, 104–113.

This article represents one of the first efforts to obtain a scientific critique of Drexler's vision of molecular nanotechnology. Garfinkel is a science writer with a triple major in chemistry, political science, and history of technology from Columbia University.

   The article is divided into four parts: an original presentation by Garfinkel, a response by Drexler, counter-response by Garfinkel, and a final rebuttal by Drexler. In the first part of the article, Garfinkel outlines Drexler's ideas about molecular nanotechnology and declares the "Chemistry Says It Can't Happen." He seems to base this conclusion on his own knowledge of chemistry and the opinion of two chemists, Rick L. Danheiser, Professor of Chemistry at MIT, and James S. Nowick, then a graduate student in organic chemistry at MIT.

Drexler's response consists of 17 sections in which he attempts to respond to each of the substantive points raised by Garfinkel. The final two parts of the article contain few new technical points about Drexler's arguments and more of an exchange, in Drexler's words, about "style, words, and feelings."

Globus, Al, et al., "NASA Applications of Molecular Nanotechnology," *The Journal of the British Interplanetary Society*, Vol. 51 (1998), 145–152.

This article explores some of the most important potential uses of molecular nanotechnology, those in the field of space research. The article is particularly useful because it provides an excellent, well-written, general introduction to the subject of molecular nanotechnology.

Goodsell, David S., "Biomolecules and Nanotechnology," *American Scientist*, May/June 2000, 230–37. Also on the Internet at http://www.amsci.org/amsci/articles/00articles/Goodsell.html.

The author explores the relationship between nanomachines that occur in living organisms and possible synthetic analogues of such machines. He reviews some of the ways in which bionanomachines can serve as models for the construction of artificial nanomachines. A tutorial on this subject presented at the Seventh Foresight Conference in 1999 is available online at http://www.sigmaxi.org/amsci/articles/00articles/Goodsell.html.

Iijima, Sumio, "Helical Microtubules of Graphitic Carbon," *Nature*, 7 November 1991, 56–58.

This classic article describes the production of carbon nanotubes (Iijima calls them "microtubules") by evaporating carbon from a carbon-arc discharge. The paper provides the first report of the existence of carbon nanotubes and shows how easily these structures can be formed.

Joy, Bill, "Why the Future Doesn't Need Us," *Wired*, April 2000. Also on the Internet at http://www.wired.com/wired/archive/8.04/joy.html.

The author, a pioneer in the computer industry, suggests that the dangers posed by some forms of modern nanotechnology, including molecular nanotechnology, are so great that scientists should consider "relinquishing" research in these fields. Joy's remarks received wide attention in the popular press and stirred up an active debate within the molecular nanotechnology community. An abbreviated list of articles and web comments arising out of Joy's article can be found on the Foresight Institute website at http://www.foresight.org/Updates/Update41/Update41.3.html#OhJoy.

Merkle, Ralph C., "Biotechnology as a Route to Nanotechnology," *Trends in Biotechnology*, July 1999, 271–74. Also online at http://www.merkle.com/papers/bionano.html.

The author explains how an improved understanding of biological systems can be used to develop nanotechnological devices and systems for the construction of molecular machines.

Merkle, Ralph C., "Computational Nanotechnology," *Nanotechnology*, Summer 1991, 134–41. Also available online at http://www.zyvex.com/nanotech/compNano.html.

A nice presentation regarding the importance of molecular modeling in the development of molecular nanotechnology, with a review of the methodologies currently available for that form of research.

Merkle, Ralph C., "Design Considerations for an Assembler," *Nanotechnology*, Summer 1996, 210–15. Also available online at http://www.zyvex.com/nanotech/nano4/merklePaper.html

This article is of special interest because it describes in technical detail the requirements that must be met for the production of a general assembler.

Merkle, Ralph C., "Molecular Manufacturing: Adding Positional Control to Chemical Synthesis," *Chemical Design Automation News*, September/October 1993, 1–8. Also available online at http://www.zyvex.com/nanotech/CDAarticle.html.

This is a technical paper that can be read at least in part by almost any educated person, describing how chemical synthesis and molecular manipulation can be combined to produce nanoscale devices.

"Nanotechnology: Special Report," *Chemical & Engineering News*, 16 October 2000.

One of the most prestigious journals in the physical sciences presents a broad and thoughtful review of developments in nanoscience and nanotechnology with a discussion of its interaction with business and government. The four major sections of the special report deal with "Building from the Bottom Up," "New Tools for Tiny Jobs," "Firms Find a New Field of Dreams," and "Crafting a National Nanotechnology Effort."

"Nanotechnology," *SciTech Magazine*, A Special Issue, Spring 1995.

SciTech is a publication of the Cornell University College of Engineering. This issue contains five articles on nanotechnology: (1) "Prometheus Returns" ("Like fire, nanotechnology will change the world"), (2) "The Promise of Nanotechnology" ("The many potential applications of nanotech"), (3) "Understanding Nature's Machines" ("Proteins as nature's evolved nanotechnology"), (4) "Atomic Reach: AFMs and STMs" ("Today's scanning microscopes may build tomorrow's nanotechnology"), and (5) "The Nanofuture at Cornell" ("A day in the life of a Cornell student in the not-too-distant future").

Overton, Rick, "Molecular Electronics Will Change Everything," *Wired*, July 2000. Also on the Internet at http://www. wired.com/wired/archive/8.07/molelectronics.html.

This is an **especially** interesting article about developments in molecular electronics at least partly because of the interviews it includes with two of the leading authorities in the field, James Tour and Mark Read.

Port, Otis, "Molecular Machines Aren't Fantasy. Just Ask the Pentagon," *Business Week*, 30 August 1999, 90–94. Also on the Internet at http://www.businessweek.com/1999/99_35/ b364407.htm

A good introduction to the subject of molecular nanotechnology by a highly respected business publication, with special emphasis on the potential applications of molecular electronics.

Stix, Gary, "Waiting for Breakthroughs," *Scientific American*, April 1996, 94–99. Also on the Internet at http://www.sciam.com/ 0496issue/0496stix.html.

This article purports to report on the status of molecular nanotechnology and its prospects for the future, but ends up becoming one of the most controversial commentaries written to date on the subject. The Foresight Institute and some of its senior officers argue that the article is strongly biased and inaccurate. The acrimonious debate between the journal and those who support molecular nanotechnology continues in various forms for almost a year. The details of the debate are reviewed on a special Foresight webpage (http://www.foresight.org/SciAmDebate/SciAmOverview.html) and are described in the lead article of *Foresight Update* #25 (http://www.foresight.org/ Updates/Update25/).

Stix, Gary, et al. "Nanotech," *Scientific American*, September 2001. This special issue contains about a dozen articles and editorials dealing with many aspects of the nanoscale revolution, including "Machine-Phase Nanotechnology" (K. Eric Drexler), "The Art of Building Small" (George M. Whitesides and J. Christopher Love), "The Incredible Shrinking Circuit" (Charles M. Lieber), "Of Chemistry, Love and Nanobots" (Richard E. Smalley), "Nanobot Construction Crews" (Steven Ashley), and "Less is More in Medicine" (A. Paul Alivisatos). The issue contains some strong opinions as to why it is physically impossible ever to construct Drexlerian nanodevices, a position that set off another vigorous debate between those who agree with these authors and those who view such devices as real long-term possibilities.

"Taking the Initiative," *Science*, A Special Issue, 24 November 2000. *Science* magazine published its first special issue on nanotechnology on November 29, 1991. A decade later, it published its second special issue on the topic. Among the topics covered in the issue are "Is Nanotechnology Dangerous?," "Cantilever Tales," "Nano-Manipulator Lets Chemists go Mano a Mano with Molecules," "Strange Behavior in One Dimension," "Nanoelectromechanical Systems," "Powering an Inorganic Nanodevice with a Biomolecular Motor," "Atom-Scale Research Gets Real," and "Coaxing Molecular Devices to Build Themselves." Except for the research reports, most of these articles are written clearly enough for the layperson to read.

Reports

Amato, I. *Nanotechnology: Shaping the World Atom by Atom.* Washington, DC: National Science and Technology Council, September 1999. The report is available online at http://itri.loyola.edu/nano/IWBN.Public.Brochure/ This short booklet provides a general overview for the layperson about the subject of nanotechnology, with "top-down" and "bottom-up" approaches described and distinguished from each other. The report is an excellent, easy-to-read, comprehensive review of the origins, methods, and promises of nanotechnology.

Ellenbogen, James C., and J. Christopher Love, *Architectures for Molecular Electronic Computers: 1. Logic Structures and an Adder Built from Molecular Electronic Diodes.* McLean, VA: The Mitre Corporation, July 1999. Document: MP 98W0000183. The report is available online at http://www.mitre.org/technology/nanotech/

This so-called pink book is a technical report showing how developments in molecular electronics thus far (the invention of molecular wires and nanoscale diodes and switches) can be combined to produce molecular-scale electronic digital computer logic. For the general reader, the report is extremely valuable because it provides a clear and complete explanation of the nature of the fundamental units of a molecular electronics (molecular wires, diodes, and switches) and summarizes the research on these objects as of late 1999.

Nelson, Max, and Calvin Shipbaugh, *The Potential for Molecular Manufacturing.* Santa Monica, CA: The RAND Corporation, 1995.

The RAND Corporation was created as a "think tank" to deal with defense-related issues. It has become one of the nation's most highly respected organizations for the analysis of important scientific, technological, economic, and social issues. In this report, it analyzes the prospects for the development of molecular nanotechnology research in the near future.

The major portion of the report is contained in Chapters 2 through 5, dealing with "Trends and Goals," "Developing Incremental Checkpoints," "National and International Research Efforts," and "Conclusions and Recommendations." A valuable appendix lists research centers at which programs in molecular nanotechnology currently are operating. This list is now somewhat out of date.

The report's primary recommendation is that a working group of researchers from biotechnology, chemistry, computer science, electrical engineering, materials science, mechanical engineering, and physics be formed to determine whether research in molecular manufacturing should be allowed to develop on a *laissez-faire* basis, or whether some type of coordination should begin to occur among researchers in the field.

The report can be ordered by telephone (310-451-7002), by fax (310-451-6915), by e-mail (order@rand.org), or by regular mail from Rand.

President's Committee of Advisors on Science and Technology. *National Nanotechnology Initiative: The Initiative and Its Implementation Plan.* Washington, DC: National Science and Technology Council, July 2000.

This report is an extension of the report, *National Nanotechnology Initiative: Leading to the Next Industrial Revolution*, which was included as a supplement to President Clinton's Fiscal Year 2001 Budget Report. This 140-page document provides a comprehensive introduction to the subject of nanotechnology, both "top-down" and "bottom-up" approaches; describes its potential impact on human civilization in general and American society in particular; outlines the goals and strategies of the National Nanotechnology Initiative (NNI); defines the roles of the seven federal agencies involved (Departments of Commerce, Defense, Energy, and Transportation; the National Aeronautics and Space Administration; the National Institutes of Health; and the National Science Foundation); outlines the programs to be funded through NNI; and includes an appendix with 14 examples of current and ongoing research programs in nanotechnology. The report is almost certainly the best single document outlining the federal government's perceived role in the development of nanotechnology in the period 2001–2006.

Rocco, M. C., R. S. Williams, and P. Alivisatos, eds. *Nanotechnology Research Directions: IWGN Workshop Report.* Washington, DC: National Science and Technology Council, September 1999. The report also is available online at http://itri.loyola.edu/nano/IWGN. Research.Directions/

This report summarizes the views of experts from government, academia, and the private sector expressed at a workshop sponsored by the Interagency Working Group on NanoScience, Engineering and Technology on January 27–29, 1999. These experts agreed that the benefits of nanotechnology can best be realized through a cooperative effort involving universities, industry, government agencies at all levels, and national laboratories. The report describes in detail the role that each of these entities could play in a national nanotechnology initiative and recommends levels of funding necessary to obtain the stated goals.

Siegel, R. W., E. Hu, and M. C. Rocco, eds. *Nanostructure Science and Technology: A Worldwide Study.* Washington, DC: National Science and Technology Council, September 1999. The report also was published by Kluwer Academic Publishers, Dordrecht, The Netherlands, 1999 (ISBN 0-7923-5854-6) and is available online at http://www.itri.loyola.edu/nano/toc.htm.

This report was prepared under the guidance of the National Science and Technology Council and the Interagency Working Group on NanoScience, Engineering and Technology, whose representatives come from eight federal agencies (Office of Naval Research, National Science Foundation, National Institutes of Health, National Institute of Standards and Technology, Office of Scientific Research, Department of Commerce, Department of Defense, and National Aeronautics and Space Administration). The report attempts to assess progress in research on "top-down" and "bottom-up" approaches throughout the world. Six of the report's eight chapters deal with specific subject matter areas, such as synthesis and assembly, dispersions and coatings, and functional nanoscale devices. Chapter 1 provides a general introduction and overview to the topic, whereas Chapter 8 offers an overall analysis and summary of the status of research programs in various countries. The appendices provides site reports from Europe, Japan, and Taiwan and a useful glossary of terms.

Siegel, R. W., E. Hu, and M. C. Rocco, eds. *R&D Status and Trends in Nanoparticles, Nanostructure Materials, and Nanodevices in the United States.* This report also available online at http://itri.loyola.edu/nano/US.Review/

This document is a report of a workshop sponsored by the World Technology Evaluation Center at Loyola College, Maryland, held on May 8–9, 1997. Following two introductory and overview chapters, Chapter 3 summarizes current U.S. government activities and interests in nanotechnology. It focuses on the work of about a dozen government agencies, including the National Science Foundation, Office of Naval Research, Department of Energy, Army Research Office, National Aeronautics and Space Administration, National Institutes of Health, and National Institute of Standards and Technology. Chapters 4 through 9 describe research and development programs in the fields of synthesis and assembly: bulk behavior; dispersions and coatings; high surface area materials; devices; and biological, carbon, and theory issues.

Chapter Twelve

Nonprint Resources

The pace of research in nanotechnology is such that electronic sources have become at least as important as print resources in reporting and commenting on discussion in the field. As with most areas of the Internet, websites devoted to nanotechnology tend to appear, disappear, and change at a significant rate. The sites listed here are thought to be among those most likely to endure and were accessed last on March 5, 2001. The Internet resources listed have been divided into two sections. The first consists of websites devoted largely or exclusively to molecular nanotechnology. The second contains specific articles about molecular nanotechnology of interest to the general reader. The final sections of the chapter list a few videos and other types of material relating to the topic of molecular nanotechnology.

Websites

Brad Hein's Nanotechnology Site
http://www.nanosite.net/

This site is one of the oldest, most complete, and most helpful resources on the Internet for information about molecular nanotechnology. The site contains up-to-date news about developments in the field, a message board and forum, and a superb guide to nanotechnology that provides an extensive background to the field, information on research and researchers, and a glossary.

Breakthrough!

http://www.lucifer.com/~sean/BT/

The webmaster for Breakthrough! describes his site as "unabashed technophilia, a newsletter on important developments in science and technology." Articles are added to the website on an irregular basis, whenever there is sufficient news of interest to make changes. A separate category for nanotechnology is provided in each edition of the newsletter, and in many cases, the research reported involves "bottom-up" developments. Some articles included on the page include "Micro-Guitar," "Smallest Molecular Motor Found in Cell," "NMR Microscope Measures Nanoforce," and "Nanotubes and Nanoprobes."

Bucky News Service

http://www.physik.uni-oldenburg.de/bucky/htmls/ and
http://physnet.uni-oldenburg.de/PhysNet/physdoc.html

Bucky News Service originally was created to provide a weekly listing of latest reports on carbon-60 and related structures, such as the fullerenes. This service is of special value and importance to those interested in molecular nanotechnology because the fullerenes are and probably will continue to be among the most important materials used in the construction of nanoscale devices. Bucky News Service discontinued operation in 1994, although the documents it contained are still available at the above-listed address. Documents dealing with fullerenes published since 1994 now can be accessed through a new website, PhysDoc, whose address is given above.

Daily Diffs

http://www.dailydiffs.com/dop000sp.htm

This website offers a hodgepodge of links ranging from discussions of molecular information theory and molecular simulations to job opportunities with Zyvex, Carl Zeiss, and other companies.

Foresight Debate with *Scientific American*

http://www.foresight.org/SciAmDebate/SciAmOverview.html

This website reviews the year-long debate between FI and *Scientific American* editors arising out of an article written in the April 1996 issue of the magazine about molecular nanotechnology. The website is useful for those interested in molecular nanotechnology because it allows one to have a better understanding of the kinds of criticisms raised against Drexler's ideas and the responses provided to those criticisms.

Fusion Anomaly: Nanotechnology

http://www.dromo.com/fusionanomaly/nanotechnology

This is an off-beat website providing news about developments in the field of molecular nanotechnology and links to news groups, chat groups, and other resources.

IEEE Nanotechnology Virtual Community

http://ieeenano.mindcruiser.com

This website was established to provide a way by which IEEE (Institute of Electrical and Electronic Engineers) could make use of its basic competencies, associated technologies, and new forms of communication to improve and strengthen nanoscale research and development and to make learning resources in the field available to researchers, educators, and industry. The website provides a resource through which interested individuals can have access to reports, access the IEEE's nanotechnology library, and discuss issues related to nanotechnology.

Jim's Molecular Nanotechnology Web

http://www.halcyon.com/nanojbl/

This website provides basic information about molecular nanotechnology with some useful links to other resources. The site is limited in its usefulness because it has not been updated since late 1998.

kr's nano page

http://www.n-a-n-o.com/nano/

This is the website for Markus Krummenacker, one of the long-time proponents of research in molecular nanotechnology. It contains an important article written by Krummenacker, "Steps Toward Molecular Manufacturing," in 1994 for the now-defunct journal *Chemical Design Automation News.*

LookSmart: Nanotechnology

http://www.looksmart.com/eus1/eus317836/eus317914/eus53777 /eus554945/

This LookSmart category provides subsections on "Nanotechnology Images, Videos and Simulations"; "Molecular Engineering and Nanotechnology Organizations"; "Nanotechnology Laboratories and Institutes"; "Molecular Engineering and Nanotechnology Companies"; "Molecular Engineering and Nanotechnology Departments"; "Molecular Engineering and Nanotechnology Guides"; "Molecular Engineering and Nanotechnology Articles"; "Molecular Engineering and Nanotechnology Books"; "Molecular Engineering and Nanotechnology Reports"; and "Molecular Engineering and Nanotechnology Periodicals."

MathMol
http://www.nyu.edu/pages/mathmol/

MathMol (Mathematics and Molecules) is intended as a starting point for anyone interested in molecular modeling. The site provides a broad and comprehensive overview of the topic, with detailed information on public domain software, a library of some three-dimensional molecular structures, activities appropriate for students in the K–12 grades, and experimental versions of two hypermedia textbooks designed for elementary and middle school students.

Molecular Information Theory and the Theory of Molecular Machines
http://www-lecb.ncifcrf.gov/~toms/

This website is maintained by Tom Schneider of the Laboratory of Computational and Experimental Biology at the National Institutes of Health in Frederick, Maryland. The page contains a vast array of topics, including a glossary for molecular information theory, information on molecular machines, books and Internet resources on information theory, and programming tools.

Nanodot: News and Discussion of Coming Technologies
http://nanodot.org/

Nanodot is a Foresight Institute website launched in June 2000 with the goal of providing the latest news on molecular nanotechnology and other emerging technologies. Information on the website usually is updated many times each day. This site is one of the best sources of information on latest developments in molecular nanotechnology and related fields of science, technology, and engineering.

Nanodot has been constructed on the model of the Slashdot website, in which a visitor can select a topic and follow it as he or she wishes. By choosing a user name, the visitor also can post comments or submit stories to the site. Some of the nearly three dozen topics currently available on nanodot include Future War, Investment and Entrepreneuring, MEMS, New Institutions, Economics, Genetic Science, Robotics, Space, Biostasis, Memetics, and Future Medicine.

The Nanoelectronics & Nanocomputing Home Page
http://www.mitre.org/research/nanotech/

This website is maintained by members of the Nanosystems Group at the MITRE Corporation. It contains an abundance of useful information about nanotechnology, in general, and nanoelectronics and nanocomputing, in particular. The section of the website called "Gateway to the Nanocosm," for example, contains articles on "Introduction to Nanotechnology and Nanocomputers," "Technologies for Nanocomputing," "Technologies for Nanoelectronics," "Technical Articles about Nanoelectronics," "Who's Who in Nanoelectronics," and "World-wide Nanoelectronics Research."

Two other large sections are devoted to information about "Where to Learn More" and "Status and Prospects: What's Been Accomplished and What Lies Ahead?" The latter section is especially useful because it provides information on the top 10 recent developments in nanoelectronics and the top 10 hard problems in the field.

Nanoforum

http://www.nanoforum.org

Nanoforum is administered by COST (European Cooperation in the field of Scientific and Technical Research), an organization formed in 1971 to coordinate nationally funded research in Europe. The organization currently has 33 member states, one cooperating state (Israel), and nine states with participating institutions (including the United States and Canada). More than 30,000 scientists involved in about 200 research projects report their findings to the website and draw on its resources.

The website focuses primarily on "bottom-up" research funded by individual member nations, by multinational groups within the European community (such as the European Commission, the European Space Agency, and CERN, the European Laboratory for Particle physics), and by COST itself.

Nanoinvestor News

http://www.nanoinvestornews.com

Nanoinvestor News is devoted primarily to news about investment in nanoscience research and development. The site contains a number of articles on this topic as well as information about conferences and events, as well as companies involved in nanoscale research and development. See also Nanotech News.

NanoLink

http://sunsite.nus.edu.sg/MEMEX/nanolink.html

NanoLink describes itself as a guide to key nanotechnology sites on the web. It is arguably the best resource to such sites, listing more than 50 websites whose primary focus is molecular nanotechnology.

The site originally was developed as a collaborative project between Memex Research, a research and information technology service based in Singapore, and Internet Research & Development Unit, which no longer exists. The site is hosted by SunSITE at the Computer Centre, National University of Singapore.

Among the topics covered in NanoLink websites are fullerenes and biofullerenes, molecular engineering and molecular manufacturing, molecular self-assembly, nanobiology and nanomedicine, nanochemistry, nanocomputers and nanocomputing, nanoelectronics, nanofabrication, nanophysics, quantum computing and quantum engineering, and scanning tunneling and atomic force microscopy. The site is user-friendly with a search function that allows one to search for any basic term in nanotechnology.

Nanomagazine
http://www.nanomagazine.com

Nanomagazine.com first appeared in 2001 with a format consisting almost entirely of interviews with individuals working in the field of nanotechnology. The plans expressed for the future of the website were to continue such interviews and, eventually, to publish peer-reviewed papers in the field of nanotechnology. The fields in which the website managers are interested include scientific software related to nanotechnology, nanomaterials and nanostructures, new polymers and molecules, financial aspects of nanotechnology development, self-assembly, scanning probes, and nanoelectronics and nanodevices.

Nanopath
http://websites.golden-orb.com/Nanopath/

This website was created by and is administered by Australia's Health-IT (Health Information Technology) agency to provide a means through which interested individuals will be able to explore new scientific ideas and discuss social issues related to the development of nanoscale science and technology.

Nanorevolution
http://www.nanorevolution.com/

Nanorevolution provides a message board and chat room for people interested in nanotechnology.

Nanospot: A Search Tool for Nanotechnology
http://www.nanospot.org

Nanospot is a targeted search engine, that is, a search engine that focuses on a specific topic, nanotechnology in this case. The website provides headlines and news on developments in the field, classified advertisements, and links to other web pages dealing with nanotechnology. As an option, searches of the complete web also can be conducted through this site.

Nanotech News

http://www.nanotechnews.com

The Nanotech News website contains a number of articles dealing with research and development in nanoscience and nanotechnology as well as business issues related to these fields. Some articles are written specifically for the website, while others appear as links from other sources. The website has an extensive and useful glossary of terms used in nanoscale science and technology. It also provides a number of links to other useful resources, such as books and journals in the field.

Nanotechnology

http://www.dvtech.com/pages/pages/TecNANO.htm

Nanotechnology is a subdivision of The Technology Page, a website providing broad, general information on many areas of technology. The site provides links to articles and resources in molecular nanotechnology.

Nanotechnology

http://www.nanotech.about.com

As with About.com's other sites, Nanotechnology is a conduit to many other sites dealing with this topic. As such, the site is a useful beginning point for a broad-ranged search on the subject of nanotechnology.

Nanotechnology

http://www.yahoo.com/

The Nanotechnology listing on Yahoo! is a subdivision of its Science category. The listing contains many subsections dealing with various aspects of nanotechnology, such as laboratory groups, research centers, and governmental agencies. The listing is neither comprehensive nor selective, but it does lead to a few of the important sites relating to molecular nanotechnology.

Nanotechnology

http://www.zyvex.com/nano/

This page is a part of the Zyvex company's website. Zyvex is the first commercial entity formed to conduct research on molecular nanotechnology, and the information provided on this page is among the most complete and useful available anywhere on the web.

Nanotechnology
http://dir.hotbot.lycos.com/Science/Technology/Nanotechnology/

This HotBot directory is a conduit to sites containing articles, books, conferences, funding information, and discussions of social and political implications of nanotechnology. The website is a good beginning point for a general search on topics in nanotechnology.

Nanotechnology Database
http://www.itri.loyola.edu/nanobase/

Nanotechnology Database is a comprehensive collection of resources in nanotechnology sponsored by the National Science Foundation. It is divided into five major sections, each of which is divided further into smaller subsections. The major sections and subsections are as follows: (1) Research and Development Centers on Nanotechnology: Academic, Industrial, and National Laboratories; (2) Funding and Sponsoring Agencies and Societies: Government Agencies and Professional Societies and Non-Profit Organizations; (3) Major Research Reports and Publications: Books and Periodicals; (4) Subjects: Electronics, Nanodevices, Nanostructures, Research, and Theory; and (5) Conferences.

The site contains material relating to molecular nanotechnology and more traditional "top-down" programs for extending microtechnology to the nanometer level. The links provided vary widely in quality in usefulness and do not seem to have been evaluated for content or appropriateness to any great extent. Nonetheless, the site is probably one of the most valuable and useful Internet sources on molecular nanotechnology.

Nanotechnology Institute
http://www.nanotechnologyinstitute.org

The Nanotechnology Institute is administered by ASME (American Society of Mechanical Engineers) as a clearinghouse for the association's activities in the field of nanoscale science and technology. The institute provides a range of interdisciplinary programs and activities designed to bridge science, engineering, and their applications. The major divisions of the website provide information on on-going research in the field as well as links to related websites.

Nanotechnology Papers

http://nanotech.rutgers.edu/nanotech/

Webmaster for this site is John Storrs Hall, now Research Fellow at the Institute for Molecular Manufacturing. Hall created the website while he was affiliated with Rutgers University. It contains fundamental information about molecular nanotechnology and links to many other important molecular nanotechnology websites. There is an introduction to the concept of *utility fog*, an all-purpose molecular nanotechnology material first developed by Hall in the early 1990s.

National Nanotechnology Initiative

http://www.nano.gov/

The National Nanotechnology Initiative was created in 2000 as part of President Clinton's 2001 Fiscal Year budget program. This website contains all of the relevant documents dealing with this initiative.

Nanotechnology Now

http://nanotech-now.com

Nanotechnology Now provides information about and links to a very wide variety of nanoscience- and nanotechnology-related topics, including professional societies, books and journals, a chat site, current events, news about developments in the field, conferences and other events, articles, and interviews.

Nanotech Planet

http://www.nanotech-planet.com

The focus of the Nanotech Planet website is the commercial possibilities developing as a result of research in nanoscience and nanotechnology. It carries many articles on this topic of interest to those in the business community. The website provides a membership option to individuals and companies, as well as background information on nanoscience and nanotechnology and stock reports on companies in the field.

Nano-tek

http://www.nano-tek.org

Nano-tek is a website designed by and for postdoctoral researchers and Ph.D. students in the field of nanoscale science and technology. The website contains articles written by researchers in these fields as well as a curriculum vitae data bank, links to related websites, references, a glossary, a newsletter, and archives of the sci.nanotech website. Topics covered on the website include nanomaterials, nanolithography, nanoelectronics, nanomagnetics, nanorobots, nanomedicine, nano-biotechnology, and biodevices.

News: Preparing for Nanotechnology
http://www.foresight.org/hotnews/index.html

This Foresight Institute website is subdivided into three major parts: (1) Nanotechnology Headlines, (2) What People Are Saying about Nanotechnology, and (3) News about Foresight. The site is updated regularly.

Overview of Nanotechnology
http://nanotech.rutgers.edu/nanotech/intro.html

J. Storrs Hall ("Josh") is the author of this general introduction to molecular nanotechnology. Hall draws on earlier papers by Drexler and Ralph C. Merkle for the information provided here.

Prepared Statements for House Science Committee, Subcommittee on Basic Research, June 22, 1999
http://www.house.gov/science/106_hearing.htm

This website contains prepared statements made by members of the subcommittee and witnesses appearing before the subcommittee for a hearing on "Nanotechnology: The State of Nano-Science and Its Prospects for the Next Decade." This hearing was instrumental in the development of a National Nanotechnology Initiative first funded by the federal government in 2001.

Commentaries on the hearing can also be accessed on the web at http://www.aip.org/enews/fyi/1999/fyi99.106.htm, http://www.eetimes.com/story/OEG19990625S0017, and http://www.techweb.com/wire/story/TWB19990628S0005.

Sci.nanotech Archives
http://discuss.foresight.org/critmail/nano_archives.html

This website contains an extended series of questions, answers, and comments about nanotechnology from 1988 to the present. The site is subdivided into archives from the period 1988 to 1994 and from 1995 to the present and has a monthly breakdown of correspon-dence. Nearly every imaginable topic relating to molecular nanotech-nology can be found on the site.

SPM Links
http://spm.aif.ncsu.edu/spmlinks.htm

North Carolina State University maintains a website describing its own work with scanning probe instruments and provides an excellent page of links to other institutions doing similar work and related resources in the field, such as journals and meetings.

Texas Nanotechnology Initiative
http://www.texasnano.com

Texas Nanotechnology Initiative is a consortium developed to bring together researchers and corporations interested in nanoscale science and technology. The group was organized by two officers from the Zyvex Corporation and Dr. Bill Osborne, Dean of the Engineering College at the University of Texas at Dallas. The group describes itself as a kind of chamber of commerce that focuses on nanotechnology, promoting business, education, and investment in the State of Texas.

WebRing
http://nav.webring.yahoo.com/hub?ring=nanotechnology&list/

This website provides links to about 35 other sites dealing with some relatively diverse aspects of nanotechnology.

Articles

Drexler, K. Eric, "Nanotechnology and 'Nanotechnology'," http://www.foresight.org/Updates/Update15/Update15.3.html.
The use of the term *nanotechnology* often leads to some confusion. That confusion arises from two different ways of looking at this field, from the "top-down" or the "bottom-up" research approach. In this article, Drexler reviews the problems involved in using this term and the ways in which it can be interpreted and understood.

Drexler, K. Eric, "Studying Nanotechnology," http://www.foresight.org/Updates/Briefing1.html.
Drexler originally wrote this article in 1988 in response to queries about possible careers in molecular technology. It is divided into four major sections: (1) Fields of Research, (2) Background Fields, (3) Levels of Knowledge, and (4) Modes of Learning. The article was updated in 1998 with the addition of Internet links related to the topic of careers in nanotechnology.

Globus, Al, "Molecular Nanotechnology in Aerospace: 1999," http://www.nas.nasa.gov/~globus/papers/NanoSpace1999/paper.html.
The author of this article has been interested in the applications of molecular nanotechnology for many years. This article is an excellent, detailed, and comprehensive overview of the potential for developments and progress in this field.

Merkle, Ralph C., "Molecular Manufacturing: Adding Positional Control to Chemical Synthesis," http://nano.xerox.com/nanotech/CDAarticle.html.

This article appeared originally in the September/October 1993 issue of *Chemical Design Automation News* and was revised for inclusion at this website. The article is especially useful because it describes many of the technical challenges facing workers in molecular nanotechnology, but it is written at a level that many high school students can understand. The author deals with issues such as the goals of molecular nanotechnology, the importance of diamonoid materials, and the tools currently available for the design and construction of nanoscale devices.

Merkle, Ralph C., "A New Family of Six Degree of Freedom Positional Devices," http://nano.xerox.com/nanotech/6dof.html.

This article is an Internet version of one that originally appeared in the journal *Nanotechnology*, June 1997, pages 47–52. It describes computer studies of a nanodevice that has six degrees of freedom, that is, is able to move independently in six different directions. The advantage of this device is that it has a simple design, has high stiffness and strength, and has a high flexibility of motion. As such, it possesses many of the characteristics one would expect of a potential assembler.

[Merkle, Ralph C.], "Self Replication and Nanotechnology," http://nano.xerox.com/nanotech/selfRep.html.

Self-replication is a fundamental and essential property of certain of the nanoscale devices to be developed through nanotechnology. In this article, Merkle describes some general properties of self-replicating systems and examines the technical challenges involved in constructing them.

Merkle, Ralph C., "Whither Nanotechnology?" http://itri.loyola.edu/nano/us_r_n_d/08_06.htm

This article provides a broad, general introduction to molecular nanotechnology, including goals in the field, the need for low-cost self-replicating systems, the role of molecular modeling, and the modeling of the first assembler.

Pesce, Mark, "Thinking Small," *Feed Magazine*, 8 November 1999. Available on the Internet at http://www.feedmag.com/invent/peace.html.

This long article provides a clear and comprehensive overview of the nature of molecular nanotechnology and how it will change the world as we know it.

Platt, Charles, "The Museum of Nanotechnology," http://www. wired.com/wired/scenarios/museum.html.
The author sets this piece 50 years into the future and describes "historic" steps forward in the development of molecular nano-technology, including the first biological RAN chip (2005), the first asteroid colonization and mining (2050), the first anticancer nanomachine (2030), and the first multifunction molecular manipulator (2025).

Smalley, R. E., "Nanotechnology and the Next 50 Years," http://cnst.rice.edu/dallas12/96.html.
This paper was presented on December 7, 1995 at the University of Dallas by Nobel Laureate Richard Smalley. Although the paper is now dated, it has considerable value because of the vision that Smalley presents for the ways in which carbon nanotubes may be expected to have essential roles in the development of nanomachines in the future.

Smith, Richard H., II, "Molecular Nanotechnology: Research Funding Sources," http://nanozine.com/nanofund.htm.
This is an extraordinary document that anyone interested in molecular nanotechnology should read. The author was particularly interested in discovering what role the federal government has had in the support of research on molecular nanotechnology. In the process of answering that question, he made some interesting discoveries as to the type and extent of research currently going on in the field and the federal government's role in that research. In addition, he uses the paper as a forum for asking what role the government *should* have and how satisfied the general public should be with the kind of research the government is currently supporting. He provides an interesting overview of the status and potential impacts of research in molecular nanotechnology, both from a technical and a more generally philosophical context.

Walker, John, "The Coming Revolution in Manufacturing," http://www.fourmilab.ch/autofile/www/chapter2_84.html.
Although this article originally was written in 1990, it still provides a well-organized analysis of the changes that have taken place in the

manufacturing process over time and the changes that are likely to occur in the future as the result of developments in molecular nanotechnology.

Videos

Foresight Institute Conferences
Video and audio tapes of Foresight Institute conferences are available from:
Sound Photosynthesis
PO Box 2111
Mill Valley, CA 94942-2111
Tel: (415) 383-6712
E-mail: faustin@aol.com or creon@netcom.com

Nanotechnology: Possibilities and Prospects
Frontier Research Seminars
1510B Hamilton Street
Somerset, NJ 08873
Tel: (908) 873-5374
E-mail: 71531.1617@Compuserve.com
This video is a 4-hour presentation by Dr. J. Storrs Hall in which the fundamental concepts of molecular nanotechnology are presented. The possible applications of the field are described and discussed.

Nanotechnology as a Shortcut to Space Settlement
Hughes Cormier Film & Video
924 Rue Senneterre
Sainte-Foy, Quebec, G1X 3Y3
Canada
Tel: (418) 658-1936
This is a 39-minute VHS tape of a talk given by Drexler in 1994 at the International Space Development Conference in Toronto, Canada.

Chapter Thirteen

Glossary

The field of nanotechnology draws on terminology used in physics, chemistry, engineering, mathematics, biology, and other sciences. It also has developed many of its own terms and phrases. This glossary includes the basic terminology of molecular technology and terms from other fields of science with special significance to nanotechnology.

Ab initio   Literally, "from the beginning," a term that describes a type of research in which (usually) large and complex structures are built starting out with simple, basic units.

Absolute zero   The coldest possible temperature; the point at which all atomic and molecular motion ceases. Absolute temperature is equal to 0 K (on the Kelvin scale) or $-273.16\,°C$.

Active site   The region on an enzyme molecule to which a substrate binds and at which a catalytic reaction takes place.

Adenosine triphosphate (ATP)   A molecule that consists of a nitrogen base (adenine) attached to the sugar ribose and three phosphate groups. ATP is the primary source of energy used in cellular reactions.

Allotropes   Two or more forms of the same chemical element that differ from each other in physical and chemical properties.

Amino acid   An organic compound that contains two distinctive groupings of atoms, one the carboxylic group ($—COOH$) and the other the amine group ($—NH_2$). Amino acids are the building blocks from which peptides and proteins are made.

Assembler   A molecular machine that can be programmed to build structures beginning with atoms and molecules. An assembler also can be programmed to construct a copy of itself, in which case it is called a *replicator*.

Atom   The smallest part of a chemical element, consisting of a positively charged nucleus and one or more negatively charged electrons. Atoms are the building blocks from which all materials and objects in the world are constructed.

Atomic force microscope (AFM)   A modification of the scanning tunneling microscope that maps the surface of an object by measuring the amount of force exerted by atoms and molecules on the surface on a cantilever tip on the microscope. The AFM has been used widely not only as an imaging device, but also as an instrument for manipulating atoms, molecules, and very small groups of particles.

ATP   *See* adenosine triphosphate.

ATP synthase (ATPase)   An enzyme that catalyzes the synthesis or decomposition of the compound ATP in cells.

Automated engineering   The use of computers to perform engineering design.

B-DNA   *See* DNA.

Biochemistry   The study of chemical compounds that occur in living organisms or their products.

Biomolecular   A term that refers to molecules commonly found in living organisms and their products.

Bond   *See* chemical bond.

"Bottom-up" technology   A term sometimes used to describe methods for producing objects and materials for some particular function by starting with individual atoms and molecules or very small groups of atoms and molecules and building them up to some large size or some specific shape.

Buckminsterfullerene   A name given to the allotrope of carbon discovered in 1985 that consists of 60 carbon atoms arranged in a complex sphere made of pentagons and hexagons. The C_{60} molecules are often referred to as *buckyballs*.

Buckyball   *See* buckminsterfullerene.

Bulk technology   "Top-down" technology in which matter is handled in large quantities, usually many trillions of atoms or molecules at a time.

Cantilever   A long, narrow piece of wood, metal, or some other material that projects from some given support. A diving board is a common example of a cantilever.

Carbon nanotube   A tube made entirely of carbon atoms arranged in hexagons and pentagons forming the sides of the walls of the tube. Carbon nanotubes are typically a few nanometers in diameter and up to a few micrometers in length.

Catenanes Organic compounds containing two interlocking ring structures that are not actually bonded to each other.

Cell repair machine A device designed for use in the field of nano-medicine, consisting of a nanocomputer and one or more nanodevices designed to detect, report on, or repair damage that has occurred to molecules, cells, or tissues.

Chemical bond An attractive force between two atoms that holds the atoms together strongly enough so that they are able to function as a unit. Chemical bonds vary widely in their strength, ranging from strong covalent bonds to weak van der Waal forces.

Chemosynthesis The manufacture of objects by means of chemical reactions. The term is often used to describe a common form of nano-technology in which individual atoms and molecules and groups of atoms and molecules are arranged and combine with each other in such a way as to produce a larger structure with certain desirable properties.

Chirality The tendency of a molecule to occur in two different forms that are mirror images of each other. The two forms are identical in every way—number, type, and arrangement of atoms—except that they cannot be superimposed on each other, but are precise reflections of each other.

Codon A set of three nitrogen bases in a nucleic acid molecule that codes for a particular amino acid. There are 64 different codons that can be formed from the four nitrogen bases found in DNA and RNA.

Cross-linking The process by which a chemical bond forms between two separate molecules.

Daltons The unit of mass when measuring atoms and molecules that is equal to one twelfth of the mass of an atom of carbon-12 ($1.660\ 33 \times 10^{-27}$ kg). Named after John Dalton, it is also referred to as *atomic mass unit (amu)*.

Dendrimer A polymer-like substance whose growth typically begins from some central core structure and builds outward in a manner similar to the way that a tree grows. Dendrimers exist in more than 50 different shapes and differ from polymers in that their growth can be precisely controlled and directed.

Deoxyribonucleic acid (DNA) *See* DNA.

Design ahead The use of theory, computers, and other tools with which to design devices and systems that cannot yet be constructed with existing tools. Design ahead technology is especially important in nanotechnology to study assemblers, replicators, and other nanodevices that cannot yet actually be constructed in the laboratory.

Diamonoid Having the properties of diamond. Many of the materials being considered for use in the construction of nanoscale devices would be made of carbon, having structures similar to those of diamond.

Diode An electronic device with two terminals that allows the flow of electrical current in only one direction and that can be used as a switch or rectifier in a circuit.

Dip-pen nanolithography (DPN) The term used to describe a new technology developed in the late 1990s by which the tip of an atomic force microscope is used to draw fine lines on a surface.

Disassembler A nanodevice with the capability of taking a material apart one atom or molecule at a time without actually destroying the structure being disassembled.

DNA The common abbreviation for deoxyribonucleic acid, a form of nucleic acid that occurs in all cells and carries the genetic code that directs the cell to carry out many of its essential operations. DNA is a large, complex molecule consisting of two spaghetti-like strands wrapped around each other in a geometric shape known as a *double helix*. A DNA molecule can take on various shapes, depending on its surrounding environment. The most common, right-handed form of DNA is known as *B-DNA*. A second form of the molecule, arranged as a left-handed helix, is called *Z-DNA*.

Electron cloud The space around the nucleus of an atom in which an electron has a high probability of being found. The "cloud" is not a physical reality, but the expression of a mathematical probability of an electron's being in some given region around the nucleus.

Enzyme A complex structure containing at least one protein that occurs naturally in living organisms and catalyzes chemical reactions that take place in cells.

Fluorescence The production of electromagnetic radiation of some wavelength produced when an object is struck by electromagnetic radiation of some other wavelength.

Fullerene A large molecule consisting entirely of carbon atoms arranged in hexagons, pentagons, and other geometric shapes similar to or related to those found in the buckminsterfullerene (C_{60}) molecule.

Graphene A naturally occurring form of carbon consisting of flat sheets of carbon atoms bonded to each other.

Gray goo A hypothetical substance composed of large numbers of nanosize particles capable of replicating themselves out of any materials that happen to be available. In *Engines of Creation*, K. Eric Drexler outlined the threat to human civilization that would be posed by the presence of such a material produced by the noncontrolled development of molecular nanotechnology.

Kinesin A protein capable of converting the energy stored in ATP molecules into mechanical motion. It occurs in nearly all cells and is responsible for transporting a variety of biological molecules from one place to another within the cell.

Kinetic energy   The energy that a particle or object possesses as a result of its motion. The kinetic energy of an object is equal to one half the product of its mass and its velocity squared, or: $K = \frac{1}{2}mv_2$.

Limited assembler   A type of assembler that carries instructions for certain specific operations, such as the ability to build a single type of structure. The purpose of such limited instructions is to prevent the assembler from "going out of control" and making dangerous or undesirable objects or materials.

Lithography (in electronics)   Any method by which a pattern can be etched into the surface of a material. The most advanced methods of lithography use beams of electromagnetic radiation because of the fine lines they are able to produce.

Logic gate   A component of digital circuits having two inputs and one output. Seven kinds of logic gates exist, each designed to perform some specific function on the inputs it receives. In an AND logic gate, for example, the two inputs are added in such a way that only two "true" inputs (two "1" messages) produce a "true" output (a "1" output), whereas any other combination of inputs produces a "false" ("0") output.

Mechanical computer   A computing device that performs calculations using the movement of physical objects (such as sliding bars) rather than changes in electrical current. Some of the earliest computing machines were designed as mechanical computers, although no modern commercial computer operates on this principle. In his *Engines of Creation*, Drexler suggested that nanocomputers might be built that operate on mechanical, rather than electrical, principles.

Mechanosynthesis   Another term for *molecular manufacturing*, suggesting the possibility of constructing structures by assembling them one molecule at a time.

Meme   A mental concept that, similar to a gene, has the ability to replicate and evolve. Memes are in competition with each other in the intellectual world in much the way that genes are in competition with each other in the physical world.

Memory devices   Components of a computing system capable of storing information, such as by being in a "magnetized" or "nonmagnetized" state.

MEMS   An acronym for *microelectromechanical systems*, a term that, in general, describes devices and systems in which electronic devices consisting of microsize components are interfaced with various types of mechanical systems, such as the use of nanocomputers in various kinds of sensors.

Mesotechnology   A term used to refer to materials, devices, systems, and processing at dimensions between about 100 nm and 1 μm.

Messenger RNA (mRNA)   A form of RNA synthesized in the nucleus of cells using DNA as a template. Once produced, mRNA molecules carry the genetic message stored in DNA to the ribosomes, where it serves as a guide for the synthesis of proteins.

Microelectromechanical systems *See* MEMS.

Micron A millionth (10^{-6}) of a meter.

Microtechnology A term referring to objects, devices, systems, and processes at the micrometer (micron) level.

Microtubule A hollow, cylindrical-shaped organelle that occurs in cells and has many important functions, including determining cell structures and providing a conduit for the transport of materials.

Molecular beam epitaxy (MBE) A "top-down" method of nanotechnology in which a substance is sprayed on a material to produce a coating one atom or one molecule thick.

Molecular electronics The design and construction of electronic devices consisting of single molecules or small groups of molecules.

Molecular manufacturing The process of constructing objects by assembling them molecule-by-molecule using physical and chemical methods.

Molecular modeling The use of computer programs that simulate the physical and chemical properties of atoms and molecules to design, construct, and test possible molecular structures.

Molecular fabrication The construction of objects or materials by assembling molecules one at a time.

Molecular machine Any device, either natural or synthetic, with dimensions of about 1 to 100 nm.

Molecular manipulator Any tool capable of picking up or manipulating individual or small groups of atoms and molecules. Scanning probe microscopes are examples of molecular manipulators.

Molecular nanotechnology *See* nanotechnology.

Molecular recognition A term that refers to the ability of one atom or molecule to position itself correctly with regard to a second molecule so that the two particles can bond to each other.

Molecule The smallest particle of a chemical compound. Molecules consist of two or more atoms held together by means of chemical bonds.

Moore's law An hypothesis originally made in 1965 by Gordon Moore, cofounder of the Intel Corporation, predicting the rate at which advances in electronic computing technology would develop. At first, he predicted that the number of circuits on a silicon chip would double every year. He later revised that projection to a doubling every 18 to 24 months.

Multiwalled nanotubes (MWNTs) Forms of carbon nanotubes that consist of two or more concentric tubes.

N-type dopant   A substance added to a semiconductor to provide it with an excess of electrons.

Nano-   A prefix in the SI (metric) system meaning 10^{-9}, or one billionth.

Nanocomputer   A computer whose components all have nanodimensions.

Nanoelectromechanical systems   *See* NEMS.

Nanofabrication   The design and manufacture of devices with dimensions measured in a few nanometers, generally between 1 and 100 nm.

Nanolithography   Any method for inscribing lines or patterns with dimensions of only a few nanometers on the surface of a material. New methods, different from those used with traditional forms of lithography, are being developed to produce markings with such small dimensions.

Nanomedicine   The use of assemblers, replicators, nanocomputers, and other nanodevices for diagnostic and therapeutic purposes in treating human diseases and disorders.

Nanorobot (nanobot)   An automated device, usually consisting of a nanocomputer and one or more nanomachines, to carry out some given task repeatedly and with precision.

Nanotechnology   A general form of technology dealing with objects in the low-nanometer range, generally from about 1 to 100 nm. The two general types of nanotechnology are those that involve working with bulk materials, reducing them to nanometer size ("top-down" nanotechnology) and those that involve working with individual or small groups of atoms and molecules ("bottom-up" nanotechnology).

NEMS   An acronym for *nanoelectromechanical* systems. The term is comparable to MEMS (microelectromechanical systems) in that it refers to integrated systems in which electronic devices consisting of nanoscale components are interfaced with mechanical devices of similar or larger dimensions.

Newtonian physics   A set of laws and theories based originally on the work of Sir Isaac Newton in the 17th century applicable to almost all physical phenomena that can be observed on the macroscopic level. Although Newtonian physics satisfactorily deals with almost any problem of everyday life, its laws begin to break down when applied to very small objects and events on the nanometer scale. In such cases, scientists must turn to the laws of quantum mechanics to understand physical phenomena.

Nitrogen base   As used with reference to nucleic acids, an organic compound belong to the one of two families of compounds, the purines and the pyrimidines, consisting of one or more rings of carbon atoms to which one or more nitrogen atoms are attached. Nitrogen bases are components of the nucleotides from which the nucleic acids are constructed.

Nucleic acid A large, complex, polymeric compound whose structure consists of long chains of nucleotides. The two major classes of nucleic acid are those containing the sugar deoxyribose (DNA) and those containing the sugar ribose (RNA).

Nucleotide A monomeric unit from which nucleic acids (DNA and RNA) are made. Nucleotides are combinations of (1) a nitrogen base, (2) a sugar (ribose or deoxyribose), and (3) a phosphate group.

Organic chemistry The science that deals with the chemistry of carbon-containing compounds.

P-type dopant A substance added to a semiconductor to provide it with an deficiency of electrons.

Paradigm In science, a general philosophy that sets a framework for laws, theories, experimental approaches, and other features of scientific work. Scientific paradigms largely determine the goals that scientists choose and pursue and the methods they use to achieve those goals. Overturning a scientific paradigm is generally a long and difficult task because most researchers prefer to work within a "tried-and-true" framework than accepting a revolution in the way their discipline is perceived. Two examples of paradigm overthrows were the replacement of the phlogiston theory by the theory of combustion in the 18th century and the modification of Newtonian physics by quantum mechanics in the 20th century. Some observers argue that the goals and methods of "bottom-up" research represent a paradigmatic shift from "top-down" technology.

Peptide An organic compound consisting of groups of amino acids joined to each other. Proteins are polypeptides, very large chains of such groups.

Photovoltaic device Any device by which the energy of solar radiation can be converted to electrical current.

Pi bond A covalent bond between two atoms in which the electrons forming the bond extend above and below the plane of the bond. Electrons in a pi bond are more mobile than those found in sigma bonds, a fact that can be used in the design of single-molecule electronic devices.

Piezoelectric Referring to any material from which an electrical current can be obtained when pressure is applied to the material.

Polymer A molecule consisting of many identical units, joined to each other in very long chains. The basic unit of a polymer is called a *monomer*. Many polymers consist of only one monomer, repeated hundreds or thousands of times, while other polymers are made of a small number (usually two) monomers, joined in alternating sequence in the chain.

Polypeptide A polymer made of large numbers of amino acids joined to each other.

Polyphenylene A polymer consisting of many phenylene units joined to each other in a long chain. A phenylene unit is a benzene ring lacking two hydrogen atoms.

Porphyrin ring An organic structure commonly found in living organisms, consisting of four nitrogen-containing pentagonal rings. A metallic ion is often sequestered in the middle of the four rings.

Probability cloud A method for describing the likelihood of finding any particular electron in the space surrounding the nucleus of an atom. A dense probability cloud indicates a high probability of finding the electron, and a sparse cloud indicates a low probability of finding the electron.

Protein A polypeptide found in all living organisms, where it may play any one of many different essential functions. Proteins are typically very large molecules consisting of tens or hundreds of thousands of atoms, or more.

Protein engineering The study and modification of naturally occurring protein and the design and construction of synthetic analogues of natural proteins to achieve better understanding of proteins or to make possible some useful practical function.

Proximal probe Any devices capable of detecting and manipulating individual or small groups of atoms and molecules. The scanning tunneling microscope and atomic force microscope are examples of proximal probes.

Quantum dot A collection of electrons confined within a semiconductor material, all of which occupy the same quantum state.

Quantum mechanics The modern theory of physics, developed to supplement and improve on classic physical theories, explaining the nature of matter and radiation and the relationship between the two.

Receptor molecule A molecule whose physical shape allows it to incorporate and bond to some other molecule, an act that generally results in the alteration of some function in the system to which the receptor belongs, as in a living cell.

Rectifier An electrical device that allows current to flow more easily in one direction than in the other.

Redundancy A method for using multiple copies of one or more components of a system. The presence of multiple copies means that if one part of the system fails, a copy of that part will be able to take its place.

Replicator Any device that is capable of making copies of itself. In molecular nanotechnology, replicators are, along with assemblers and nanocomputers, one of three fundamental types of machines needed for large-scale molecular manufacturing.

Ribonucleic acid (RNA) A form of nucleic acid that differs from DNA in a few important respects, such as (1) it consists of a single-stranded, rather than a double-stranded, molecule; (2) it contains the nitrogen base *uracil* rather than thymine; (3) the molecule is built with the sugar *ribose* rather than deoxyribose; and (4) it occurs in at least four different forms, known as messenger RNA (mRNA), transfer RNA (tRNA), ribosomal RNA (rRNA), and viral RNA (vRNA).

Ribosomal RNA (rRNA)   A form of RNA found in the ribosomes. Its function is not yet fully understood.

Ribosome   A cell organelle present in all cells that serves as the site for protein synthesis. mRNA molecules carry instructions from the cell nucleus to the ribosome, and tRNA molecules bring amino acids to the ribosome where they are assembled according to the instructions stored in the mRNA molecules.

Robot   A mechanical device that can be programmed to move about or perform a variety of tasks automatically or under the control of an exterior director.

Rotaxane   A large, complex structure consisting of two or more independent segments not connected directly to each other but linked through a linear piece threaded through a ring.

Scanning probe microscopes (SPMs)   A general term referring to a group of instruments that operate on some modification of the scanning tunneling microscope.

Scanning tunneling microscope (STM)   An instrument originally designed to obtain images at the atomic scale of the surface of a material but later modified to remove, transfer, and otherwise manipulate individual particles on such a surface.

Self-assembly   A technique by which a system constructs a machine or some other device without the need for any external directions or input. Self-assembly occurs when molecules with congruent geometric and chemical properties are joined to each other. Self-assembly is a common and efficient process in all living organisms and is a long-term goal of molecular nanotechnology.

Self-replication   The process by which a molecule, a machine, or some other device makes an exact copy of itself. DNA molecules routinely and efficiently undergo the process known as *replication*, in which they make exact copies of themselves.

Semiconductor   A compound or material that conducts an electrical current more efficiently than an insulator but less efficiently than a conductor. The electrical properties of semiconductors have made them the basis for much of the modern electronic industry.

Sensor   Any device that is capable of detecting any one of a number of physical or chemical stimuli, such as the presence of heat, light, sound, pressure, magnetism, or some specific chemical compound.

Sigma bond   A covalent bond joining two atoms in which the electrons that make up the bond line along the plane of the two atoms. The electrons in a sigma bond are much less mobile than those in a pi bond so that compounds in which sigma bonds predominate tend to be effective insulators.

Single-walled nanotubes (SWNTs)   Carbon nanotubes whose walls consist of a single tube, typically one atom or one molecule thick.

Smart materials   Materials containing nanocomputers and other nanodevices capable of carrying out a variety of relatively complex functions.

Substrate   (1) Any substance on which an enzyme operates or (2) any solid substance on which a coating or some other material is deposited.

Switch   In electricity, a device for changing the direction of an electrical current.

Template   A pattern or guide used in the construction of a material. In chemistry, some molecules act as templates for the manufacture of other molecules. Perhaps the best example is the use of a DNA molecule as a template for the manufacture of an RNA molecule during the process known as *transcription.*

Tensile strength   The maximum force that can be placed on an object or material before it fails, that is, before it tears or breaks apart.

"Top-down" technology   Another name for the form of technology that has been used throughout human history, in which a block of material is cut, polished, decomposed, or otherwise reduced in size or shape or taken apart chemically to produce an object of desired shape, size, or utility.

Tour wire   A nanowire made of some modification of a polyphenylene molecule.

Transcription   The process by which a set of information encoded in a DNA molecule is copied into a messenger RNA molecule, which then carries the message to ribosomes, the site of protein synthesis.

Transfer RNA (tRNA)   A form of ribonucleic acid whose role in cells is to carry amino acids to ribosomes, where they are assembled into new protein molecules.

Translation   The process by which a new protein is synthesized on a ribosome from information carried on an mRNA molecule and using amino acids brought to the ribosome by tRNA molecules.

Tribology   The study of interacting surfaces that are in relative motion to each other.

Tunneling   A process by which electrons pass from one conducting material to a second conducting material through a nonconducting material. According to the laws of classic physics, that movement of electrons should not be possible, but the laws of quantum mechanics do allow a few electrons to "escape" or "tunnel under" the energy barrier between the two conducting materials.

Uncertainty principle   A principle from the field of quantum mechanics that says that it is impossible to measure the position and the momentum of a particle with perfect precision and that improving the measurement of one of these variables inevitably diminishes the precision of the second variable.

Utility fog   A hypothetical material consisting of nanocomputers, assemblers, and other nanodevices that has the capability of changing size and shape and otherwise adjusting itself to new conditions and new demands placed on it.

Van der Waals forces   Weak forces of attraction between molecules. Van der Waals forces are weaker than hydrogen bonds and much weaker than either ionic or covalent bonds.

Z-DNA—*See* DNA

Acronyms

ADP   adenosine diphosphate

AFM   atomic force microscope (or microscopy)

AMP   adenosine monophosphate

ATP   adenosine triphosphate

DARPA   Defense Advanced Research Projects Agency

DNA   deoxyribonucleic acid

DOC   U.S. Department of Commerce

DOD   U.S. Department of Defense

DOE   U.S. Department of Energy

DOT   U.S. Department of Transportation

EPA   Environmental Protection Agency

ERC   National Science Foundation Engineering Research Center

FI   Foresight Institute

FY   fiscal year

HHS   U.S. Department of Health and Human Services

IBM-ZRL   IBM (International Business Machines) Zurich Research Laboratory

IMM   Institute for Molecular Manufacturing

IWGN   Interagency Working Group on Nanoscience, Engineering and Technology

LBNL   Lawrence Berkeley National Laboratory

LED   light-emitting diode

LLNL   Lawrence Livermore National Laboratory

MBE   molecular beam epitaxy

MRSEC   National Science Foundation Materials Research Science and Engineering Center

NASA   National Aeronautics and Space Administration

NFM   near-field microscope (or microscopy)

NIH   National Institutes of Health

NIST   National Institute of Standards and Technology

NNCO   National Nanotechnology Coordinating Office

NNI   National Nanotechnology Initiative

NSET   Subcommittee on Nanoscale Science, Engineering and Technology of the National Science and Technology Council's Committee on Technology

NSF   National Science Foundation

ONR   Office of Naval Research

OSTP   White House Office of Science and Technology Policy

PCAST   President's Committee of Advisers on Science and Technology

R&D   research and development

RNA   ribonucleic acid

SPM   scanning probe microscope (or microscopy)

STC   National Science Foundation Science and Technology Center

STM   scanning tunneling microscope (or microscopy)

Index

About the Author

DAVID E. NEWTON is an independent author. He has written over 400 books and articles in the field of mathematics, science, social issues of science and technology, and other fields. He has taught at the high school level, was Professor of Chemistry and Physics at Salem State College for 13 years, and was also Adjunct Professor at the University of San Francisco for 10 years.